P9-CDX-111

ENERGY

IN THE WORLD OF THE FUTURE

Other books by Hal Hellman
in the World of the Future series:

BIOLOGY IN THE WORLD OF THE FUTURE
THE CITY IN THE WORLD OF THE FUTURE
COMMUNICATIONS IN THE WORLD OF THE FUTURE
FEEDING THE WORLD OF THE FUTURE
TRANSPORTATION IN THE WORLD OF THE FUTURE

ENERGY

IN THE WORLD OF THE FUTURE

BY

HAL HELLMAN

M. Evans and Company, Inc.

NEW YORK, N.Y. 10017

M. Evans and Company titles are distributed in the United States by
J. B. Lippincott Comyany,
East Washington Square, Philadelphia, Pa. 19105
and in Canada by
McClelland & Stewart, Ltd.
25 Hollinger Road, Toronto 374, Ontario

Copyright © 1973 by Hal Hellman
All rights reserved under International
and Pan American Copyright Conventions
Library of Congress Catalog Card Number: 72-90980
ISBN 087131-123-2

9 8 7 6 5 4 3 2 1

94433

BELMONT COLLEGE LIBRARY

TJ
153
.H38

For the Mausners

The list of people who have contributed time, information or material to this book is a long one, and the author regrets that it is not possible to thank all of them here. He would, however, like to acknowledge the assistance provided by the following: E. M. Chemnitius of the Bergen Electric Generating Station (Public Service Electric and Gas Co.), P. E. Glaser of Arthur D. Little, Inc., M. King Hubbert of the U.S. Geological Survey, J. J. Kearney and Edward Kuhn of the Edison Electric Institute, L. J. Lawson of Lockheed Missiles and Space Co., Inc., R. H. Logue and W. E. Rosengarten of the Philadelphia Electric Co., Norman Metzger of the American Chemical Society, J. C. Rowley of the Los Alamos Scientific Laboratory, J. A. Scharff of Edmund Scientific Co., R. J. Schoeppel of Oklahoma State University, C. M. Summers of Rensselaer Polytechnic Institute, Lowell Wood of the Lawrence Livermore Laboratory, and especially D. S. Rose of M.I.T. Prof. Rose not only provided information but reviewed the manuscript as well. His comments and suggestions were very helpful.

Contents

I. PROLOGUE

ANDREW MANN FLIPPED a switch. "Well, that's it," he said to the assembled group of reporters, cameramen, and company officers, "the last major fossil-fueled furnace in the world has been shut down."

His little act of flipping that switch had just been flashed around the world. THE END OF THE FOSSIL FUEL ERA, the Worldview announcers called it. One of the reporters asked, "Does that mean we have actually run out of fossil fuels?"

"Only in a very rough sense. You may recall that, about ten years ago, the North American Energy Commission decided that the remaining coal reserves were to be used only for the manufacturing of essential items such as drugs and food—exactly as happened with oil and natural gas more than a century ago.

"We in the United States were lucky to have great reserves of coal, and so we were able to go on using fossil fuels, mainly by liquefaction and gasification of coal, long after the rest of the world phased them out. Liquefaction provided petroleum for our vehicles, and gasification provided synthetic gas to take the place of natural gas for cooking, heating, and production of electricity, as in this plant.

"But we had reached a danger point." He got up and pointed through the small window at the sky: "See that?"

"See what?" asked several voices.

"Clean air," he smiled. "The point is that while syncrude (synthetic crude oil, that is) and syngas are clean-burning, so much was still being burned that we were starting to repeat the twentieth century all over again.

"So although we still have fairly large reserves of coal, it is deemed best to use them for other purposes; and over the past decade these fuels have been phased out in a carefully regulated manner, this plant being the last of the major users.

"As you surely know, the 'medium of exchange' in the energy industry is now hydrogen, which is manufactured from water. All new vehicles and furnaces are being made to burn this fuel, and virtually all old ones have been converted, though some small amount of syngas is still used for cooking and heating. Mass transit, of course, is entirely electrical and has been for a long time.

"Any more questions?"

Another reporter spoke up. "Does this mean that the whole plant has to be closed down?"

"Oh, no. I said the fossil fuel *furnace* was shut down. The function of the furnace is simply to supply heat which is then converted to electricity by various means. This plant uses electrogasdynamics rather than the old-fashioned thermoelectric or magnetohydrodynamic methods, and so is relatively new. It wouldn't make sense to close down the whole thing. No, as you can tell—if you listen closely you'll hear a vague hum—the electrical generating portion of the plant is still operating, but the primary energy is now being taken from a geothermal well. In other plants, fusion, tidal, solar, wind, or hydroelectric power may be used, or some combination, depending on what is available.

"Any more questions? No? Okay. Thank you all for coming." Andrew waited for everyone to leave, then looked at his new isotope-powered watch. It was already 1520, well past his usual quitting time. He decided to take the company's still experimental all-electric car to get home. The electric guideway had finally been finished all the way out to his town and he was anxious to try out the system.

Andrew was one of the few lucky ones who had a private home; it was in a recountrified area, so he still needed a private vehicle for commutation. He was also lucky in that the area was within the allowed commuting distance from his office.

As he got into the Electracar, he punched out his destination on the control board and settled down for the 70 kilometer ride which, if all went well, would take just ten minutes.

The Electracar skimmed along nicely on its super-conducting magnetic cushion, but Andrew was frowning nonetheless. The sky, he noted, was still cloudy. Living in the country had its disadvantages too, he reflected. In return for the privilege of having a private home, he had to give up the convenience of city electricity. He had to depend on solar and wind power.

Usually this combination provided enough electricity and hot water for his family's needs; the problem arose when there were three or more successive days of calm, cloudy weather—which had happened twice already in the last two months. A three-day supply of energy was the most his storage system could hold.

Fortunately a breeze had sprung up, which meant that

his two auxiliary wind generators would be able to manufacture some electricity which they could store in the new high-density batteries. This way he wouldn't have to use expensive hydrogen to power his fuel cell power supply. . . .

Sometimes, and this was one of the times, he wondered if all this fussing and balancing of energy and energy substitutes was worth the effort involved. When the Manns first moved into their new home, the concentration on recycling, conservation, and juggling of energy resources was kind of fun; but that wore off pretty soon, and now he found that he often yearned to move back to an apartment and let the building engineer do the worrying. If nothing else, at least apartment dwellers had an assured supply of energy —or did until recently.

An urban fusion power plant.

For, as the car turned up Fusion Road, Andrew remembered that this was no longer true. He caught a glimpse of the handsome fusion plant located across the way from the middle school. The plant was now operating at only 5 percent of capacity. When proposed and built, it was thought to be perfectly safe and there were few objections raised to its location. Several had already been built in city centers.

But just recently one such plant, in New Philadelphia, had developed a severe and unexplained temperature rise in the power process. In the interests of safety, and until the problem was resolved, all such plants were closed or reduced to low operating levels until further notice. Of course activity was feverish in trying to find out what was wrong, and the outcry in the areas hit with rationing was loud and forceful. But the Energy Commission was adamant. Its members didn't want a repetition of the breeder plant radiation release that had had such a devastating effect on Old Denver, making it virtually unlivable for several years. So now even the great apartment complexes in this region, which once had a fully assured and adequate supply of energy, were on short energy rations.

Andrew was approaching the long glide down into the valley where he lived. He thought about the many ways that energy was now being conserved, as opposed to the way things used to be done, which made a fairly high standard of living possible for the two billion people living in United North America. A typical example was using the braking energy to generate electricity on downhill portions of his trip rather than letting it go off as heat. This didn't save a great deal of energy, but he agreed with the Energy Commission: Every little bit helps.

He wondered if it would be worth investing in one of the new all-home computers which would take care of such things as closing the draperies during cloudy periods to help keep in the heat.

His musings were interrupted by the car computer which warned him that he would soon have to switch over to battery operation and take the controls. As he guided the Electracar up to his compact five-room house, with the large glass wall on the south side for the solar heat system, he wondered: Would the time ever come again when there would be plenty of energy? He had read that back in the twentieth century energy was both abundant and cheap, and that—though he found this hard to believe—people rarely even thought about the subject.

1

Energy is Everything

THERE IS PROBABLY no other word in the language of man which can be defined so simply as *energy*, yet which has such a diversity and depth of meaning. Webster's dictionary calls it the "capacity for performing work." Even more simply: Energy is what makes things happen.

But think about the implications of these definitions: There is nothing you eat, drink, ride, wear, sit on, sleep in, lift up, put down, or play with that does not somehow require the expenditure of energy.

On the other hand energy, once man learned how to put it to work for him, has become his genie. For one thing, it can, like a genie, assume any one of a number of different forms. But more important, it can make strange and wonderful things happen. There is hardly anything anyone can ask a genie to do which our modern-day machines cannot accomplish, including moving mountains. Perhaps we cannot get from here to there instantly (yet), but there are supersonic jets which can zip from New York to San Francisco in an hour and a half. And man has walked on the moon, which I would guess even most genies would hesitate to try.

The real magic, however, lies in what our energy-intensive civilization has made possible for the average man. It would never have occurred to Aladdin to ask his genie to provide an individual genie for everyone in his country. Yet that is exactly what man's conquest of energy has made possible. One man, using only his own energy, can produce a steady output of about 1/10 of a horsepower (hp)* during an 8-hour workday. Nationwide, our energy production averages to about 15 hp per person. The result, as you can see, is that each of us has the equivalent of some 150 slaves (15 ÷ 1/10) working steadily for him or her every day.

Oil and gas supply 75 percent of our energy, coal only about 20 percent, and hydroelectric power about 4 percent. Nuclear power still supplies less than 1 percent. Complete figures (as of 1970) are given below.

Petroleum	43.0%
Natural gas	32.8
Coal	20.1
Hydropower	3.8
Nuclear	.3
	100.0

Energy, Force, Work, Power

At this point it might be a good idea to look more carefully at the terms *energy, work, force,* and *power.* Although often used interchangeably, they have very specific and different meanings to the scientist.

Probably the simplest and easiest to understand is force,

* See Appendix for list of conversion factors.

mainly because often it is directly perceivable by our senses. A push or a pull is a force. It can be a contact force, as when you push a couch across a room. But the effort need not only be muscular or even machine-related. Water in a can exerts a force on the bottom and sides of the can. If a dam weakens, the force of the water behind it will become greater than that which the dam can exert in holding back the water, and the water will break through.

Direct physical contact is not always necessary. There are also forces that can act through empty space. These are action-at-a-distance forces; electric and magnetic forces are examples, and are basic to the operation of the machines that have moved us into the age of technology.

Usually when we apply a force we accomplish something, as when we push a stalled car (say a Pinto) to get it moving; sometimes we apply a force and nothing happens, as when the car happens to be a Cadillac. In the first case we have performed work. In the second we have not. The point is, no matter how much energy you expend, if nothing happens you have not performed any work–in the scientific sense. Thus work is, in this meaning, the application of a force through a distance; or, alternatively, the application of effort to accomplish a task.

In a sense, work is energy that has been put to use. Gasoline can be made to do work, so it contains (or, some would say, *is*) energy. So is food, or electricity, or heat. Energy can be called the ability to do work.

The final term is power, which is defined as the rate at which work is being done (or that energy is expended). A powerful engine is one that can convert a large amount of energy to work in a short time. Put another way, power measures energy used per unit time. This will be particularly important when we talk about electricity and storage of energy.

In these terms, we see that a machine is merely a device that makes it possible for us to put energy to work. An electric motor puts electricity to work, while an internal combustion engine puts gasoline to work. And living things, it turns out, do exactly the same thing. In order to exist they too must utilize some form of energy. Plants use the energy of the sun, and animals use the energy stored up by plants (or other animals).

Power output of basic machines has climbed 10,000 times since 1750.

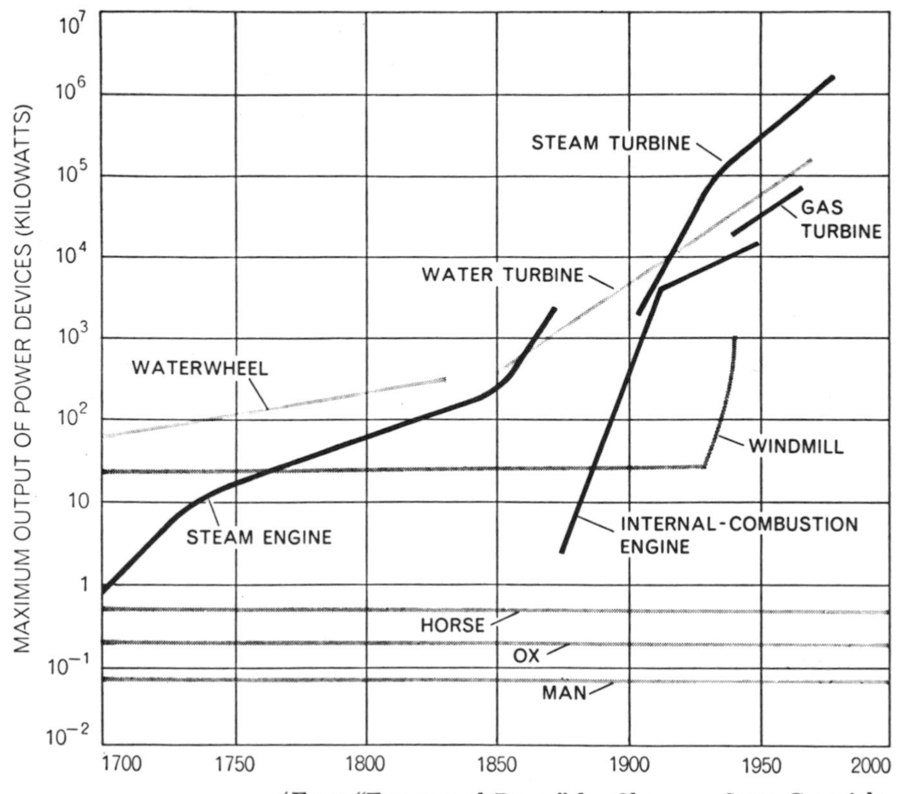

(From "Energy and Power" by Chauncey Starr. Copyright © September 1971 by Scientific American, Inc. All rights reserved.)

Forms of Energy

In general, the many kinds of energy can be categorized as follows:

• Mechanical energy. This can be energy of motion, as in a swing, a hit baseball, or the wrecker's cast-iron ball; or it can be stored energy, as in a wound-up clock spring. Another example of stored energy is the water behind a dam which, when channeled, runs through turbines and generates electricity.

• Heat. In a steam engine, fuel is burned to produce heat.* This is often used to convert water to steam, which in turn is used to perform some mechanical activity.

• Chemical energy. Wood, coal, oil, gas, and food all provide chemical energy—the energy locked up in chemical molecules. When we burn something, the light and heat come not from the destruction or disappearance of the fuel, but from chemical bonds which are being broken. If we were to weigh the products of combustion (burning), we would find that we have changed matter, not lost it.

• Light. It is the energy sunlight supplies to all the plants on earth that enables them to perform the magic called photosynthesis. All green plants depend directly on the light energy of the sun for their energy needs; and animals, ultimately, depend on plants for theirs. In addition to light, there are also X-, infrared, ultraviolet, and gamma rays; all are forms of radiant (acting-at-a-distance) energy. Thus they exist only while in motion, i.e., in traveling outward (radiating) from a source.

* Heat is different from temperature. Heat is a quantity; a certain amount of heat is needed to bring a quart of water from freezing to boiling. Heat passes from hotter to cooler bodies. Temperature, on the other hand, is a measure of hotness, or level of heat, relative to some standard such as the temperature of freezing water.

• Sound. Sound is also a form of energy. Like light, sound travels in waves; but sound waves are of a different form and need a material medium (air, water, etc.) through which to travel. High-frequency sound waves (phonons) have already been put to work in various ways, as in cleaning of instruments, and medical diagnosis and treatment.

• Electrical energy. Although electricity is generally thought of simply as a movement of electrons, it is concerned also with both electrical and magnetic forces, and interactions of the two. Of all the forms of energy, electricity is surely the most versatile, and so far the cleanest and easiest to use.

• Nuclear energy. This is the newest of all the energy forms that man has discovered and put to work. It is the most powerful form he has found and put to use so far.

The term *energy* has many implications and ramifications. Nevertheless, the basic theme of this book is easily and quickly stated:

1) Energy is the bedrock of modern civilization.
2) Energy requirements throughout the world are going to increase enormously, at least in the next few decades.
3) We must pay a price for energy production, not only in terms of money, but in the degradation of our environment.
4) New technology will supply some of the answers. But our capacity to supply enough clean energy to keep the world running will be strained considerably and will require the services not only of scientists and engineers, but of economists, politicians, businessmen, and even public relations experts (to convince us to hold down consumption.)

To begin with, let us look more specifically at the requirements that must be satisfied in coming years.

II. TODAY'S ENERGY

2

More Power to You

JUST AS THERE ARE a number of forms of energy, there are a number of ways of expressing this energy. We shall have to sort them out if we are to make any headway.

An adult, as we saw in the last chapter, can work at an average energy equivalent (or rate) of about 1/10 horsepower (hp). In the same way, when you turn on a 100-watt bulb, you are using electricity at a rate of 100 watts. At the end of an hour you will have used 100 watt-hours. This is the same as .1 kilowatt-hour or .1 kw-hr. Since 1 hp equals about .75 kw, a man can put out an electrical energy equivalent of some .075 kw, or one 75-watt bulb!

Multiplied by 8 hours, the person can produce a total of about .6 kw-hr, worth only a few pennies!

If, instead of electricity or mechanical work, our adult directed his energies directly into heat, we would need yet another measurement. The most common is that of the British thermal unit, or Btu, which is the amount of heat needed to raise the temperature of 1 pound of water 1 degree Fahrenheit. (About 150 Btus are needed to bring 1 pint of water to boil from room temperature.) There are roughly 3412 Btus to one kw-hr. A person's work output in heat is therefore .7 kw-hr × 3412 Btu/kw-hr, or about 2400 Btus in an 8-hour day.

In dealing with the energy requirements of a country like the United States, the Btu is far too small a unit, and so we find the units given in millions or even trillions of Btus.

Although the fuels consumed vary, each fuel has an energy equivalent in Btus.

Fuel	*Common Measure*	*Btus*
Crude oil	Barrel (Bbl.)	5,800,000
Natural gas	Cubic Foot (CF)	1,032
Coal	Ton	25,000,000 (approx.)
Electricity	Kilowatt-Hour	3,412

Putting all of this together, we come up with a single, total figure for energy consumption in the United States, which turns out to be about 70,000 trillion (70×10^{15}) Btus. Although this is clearly a large number, it probably doesn't have much meaning to the average person. It might help to point out that included in that total are some 5.5 billion barrels of oil, 511 million tons of coal, and 22 trillion cubic feet of natural gas.

The total energy consumption figures show that our energy consumption has been rising at a rate of 5 percent/year (more than 4 times faster than population growth.) The United States, with some 6 percent of the world's population, now consumes about one-third of the world's energy.

If the rest of the world should bring itself up to U.S. rates of consumption, then world energy usage would leap to 10 times the present rate.

Today 98.7 percent of our energy flow comes from mined resources which are rapidly being exhausted. Clearly this

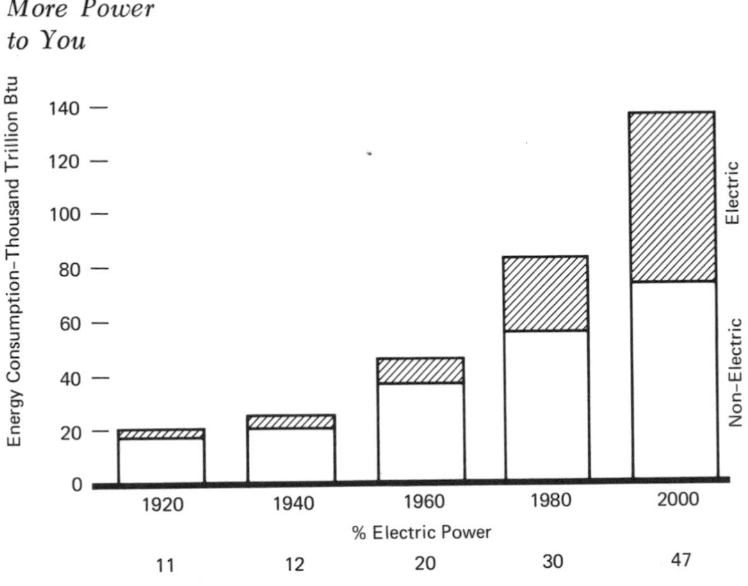

U.S. energy consumption 1920-2000.

cannot continue indefinitely. Let us see how our fuels are being used.

Where Our Fuels Are Being Used

As shown in the following table, industry is presently the largest user of energy in the United States. Direct heat or steam may be used to melt, boil, mix, separate, or otherwise transform materials. Electric utilities (generation of electricity) comes next, followed by transportation, residential and commercial uses, and, finally, nonenergy and miscellaneous. The last entry includes use of fuels as raw materials for manufacturing purposes, as in the production of dyes, plastics, paints, drugs, fertilizers, perfumes, flavors, asphalt, lubricants, and hundreds of other products.

Percentage Shares of U.S. Energy Consumption

Sector	1970	1985
Industrial	26.2	19.7
Electric Utilities	24.6	35.5
Transportation	24.0	22.6
Residential/Commercial	19.2	15.0
Nonenergy and Miscellaneous	6.0	7.2
	100.0	100.0

(Source: National Petroleum Council)

Nonenergy use of fuels is growing and, one day, it may be that this will be the *only* use permitted of fossil fuels, along with some sort of petroleum, or petroleum-based, food. A process is already in use that produces edible protein from microorganisms fed petroleum wastes. It may be that if we run short of farmland, some sort of petroleum food will have to be used in the future.

Electricity occupies a unique position in our table. While it is a consumer, it is also a producer, which sometimes makes for considerable confusion. To help keep the categories separate, it is sometimes said that electric utility plants are *users* of primary energy sources (coal, oil, etc.), and *suppliers* of secondary energy, i.e., electricity. For example, the following table shows the breakdown of fuels used in electrical generating plants, as of 1970. It should be noted that almost half of all electric power generated is used by industry.

Fuel	% of Total (approx.)
Coal	45
Natural gas	24
Hydropower	16
Petroleum	13
Nuclear	2
	100

(Source: U.S. Bureau of Mines)

Because electricity is the cleanest "fuel" we have so far, because of its great flexibility (can be used in so many ways), and because of its convenience (just plug in an appliance), it has been the fastest growing energy source of all. At present rates of growth, electricity, now providing one-quarter of our secondary energy, may provide half by the end of the century. This is shown clearly in the accompanying chart. If all goes well with nuclear power, it could go higher.

Total U.S. energy consumption by consuming sectors.

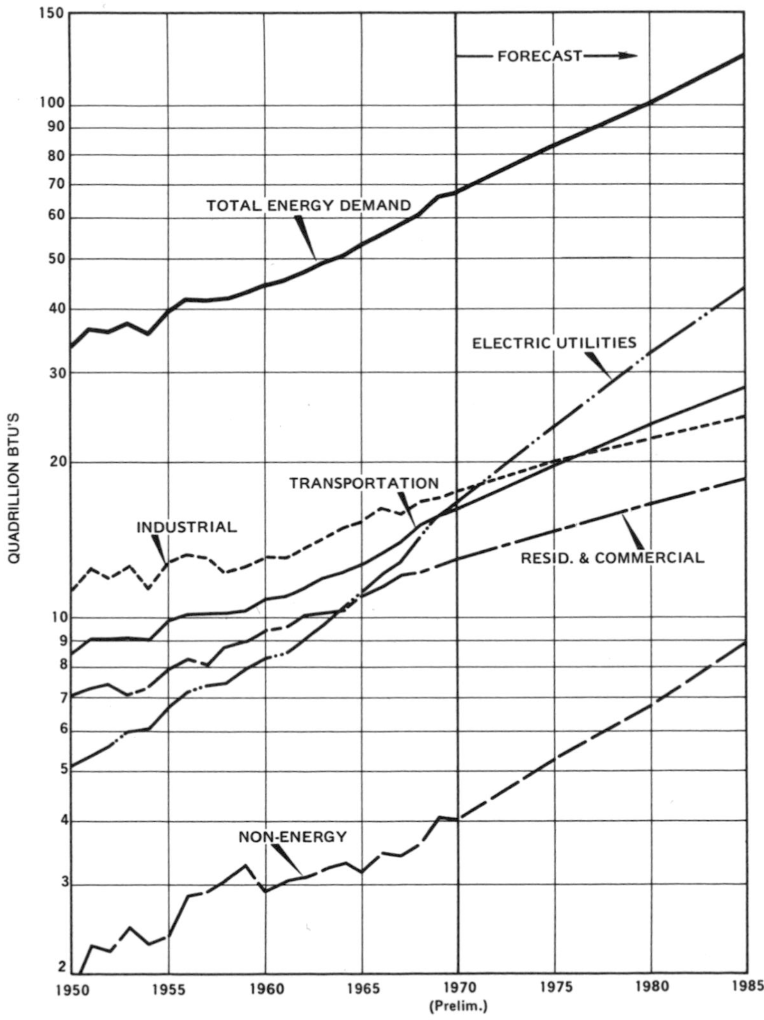

As of 1968, more than 90 percent of American homes were wired for electric power, with as many as 40 electric appliances in each, converting electric energy to thermal, acoustical, mechanical, or luminous form.

In 1971 the nation's electric utilities produced a record 1.6×10^{12} (1.6 trillion) kw-hrs. The average American household consumed 7,000 kw-hrs.

It should be noted that both of these figures are averages. There are peak times and there are periods of low demand. Thus a given generating plant at a given time may not be operating at full power, or indeed not at all. One of the important factors in the electricity energy question is that, in general, there is no satisfactory way to store electricity, so it must be produced as needed. The difference between output at any given time and total capacity is called "reserve capacity."

The electric industry believes that a reserve capacity of 20 percent is a minimum requirement, to take care of breakdowns, prolonged heat spells (which both adds to air conditioning loads and does not give overheated circuits a chance to cool off), uneven distribution, etc. The average reserve margin nationwide in the summer of 1972 was something like 17 percent, which is not bad, and was an improvement over the 1971 figure of 15.3 percent.

But the figure varies for different regions of the country, which means that some have lower reserve margins. New York, for example, with a reserve of 15 percent, was expected to have trouble, and did. A large section of Brooklyn and a small part of Queens in New York City were without power for periods up to several days during a prolonged hot spell. To help them scrape through a difficult summer, electric utility managers had to resort to voltage reductions or "brownouts" (which dim lights and cut efficiency of air conditioners) and appeals to the public to cut consumption.

Yet the Federal Power Commission estimates that demand for electricity in the next two decades will require a quadrupling of capacity! Some of this will come from increased population. But Charles Luce, chairman of Consolidated Edison in New York City, points out that a new 1000 megawatt plant (1000 million watts) will have to come on line every two years through 1980 to meet increased requirements in its area—an area in which little population growth is expected.

In other words, the major factors for the increase lie elsewhere, mainly in greater use of electric heating and air conditioning.* Other areas are increased use of electricity in industrial conversion industries, such as steel, petroleum, and aluminum, and both greater use of and "upgraded" electric appliances in the home, such as self-cleaning ovens that rely on extremely high temperatures to burn away any residue.

As an example of the kind of problems we face, consider the following. Michael McCloskey, executive director of the Sierra Club (an environmental and conservationist organization), tells us: "It has been calculated that even with large, 1000 megawatt power plants each requiring an area of only 1000 feet on a side, in less than 20 doublings—less than 200 years—*all* the available land space in the United States would be occupied by such plants!"

Clearly this is ridiculous. And yet, what is going to change? So many things point to higher consumption levels, including the call for pure air and water, and new methods of producing and preparing foods. Brilliant lighting of streets in high crime areas seems to help prevent crime, but adds to the

* Electric utilities report that as much as 35 percent of the peak summer load can be attributed to air conditioning and other weather-related needs.

electrical load. Television requires more energy than radio, and color television more than black and white. Night ball games and calls for increased use of rapid transit facilities are other examples of higher energy use.

Energy Crisis?

In 1965 a power failure in the Northeast threw 30 million persons into darkness. Thousands were caught between stops in elevators and subways. There were a number of deaths due to fright or accidents. And the inconvenience was monumental, especially in large cities like New York. Some hospitals had emergency power supplies; some did not. Traffic, unregulated by traffic lights, slowed to a crawl or blocked up altogether. This cut seriously into movement of fire engines, ambulances, and other emergency vehicles.

As a demonstration of what power means to a city, it was spectacular. As an indication of what the future may yet hold in store, it was ominous.

Although the effect of the energy shortage or crisis has been most obvious in the power field, and has stemmed largely from weaknesses in the electrical generating capacity, the future problem is mainly one of present and impending fuel shortages. In later chapters we look into the question of whether these are real shortages or, as some insist, artificial, brought on by too rapid an institution of clean-fuel and clean-power requirements, or, perhaps, by design, on the part of fuel companies. We also consider some possible alternatives. Before we do that, however, it would be well for us to discuss some of the characteristics of energy conversion.

3

Energy Conversion

A SKILLFUL BLACKSMITH can heat a piece of iron red-hot simply by hammering on it. Similarly, rooms were sometimes heated in past years by using some form of motive power (such as water power) to cause one iron plate to rotate upon another; the friction between them created heat which provided the desired warmth.

In both cases mechanical energy is being converted into heat. Indeed, these are examples of a general "law" stating that energy can neither be created nor destroyed, but can only be converted from one form to another. This law is sometimes called the first law of thermodynamics,* and sometimes the law of conservation of energy. In principle, all forms of energy can be converted to all other forms.

In other words, *use* of energy actually means conversion from one form to another, and our engines and appliances can be thought of as energy converters. The conversions

* Thermodynamics is the theory of the relations between heat and mechanical energy.

involved in a typical steam-electric plant are shown in the accompanying diagram.

Steam-electric plant conversions.

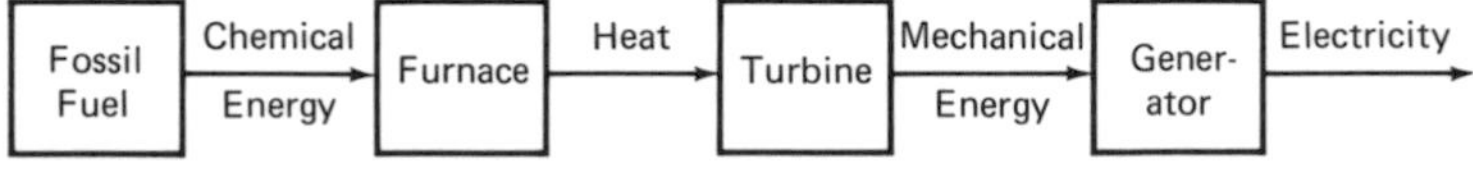

Unfortunately, in each of these conversions a certain amount of the original energy stored in the fuel being burned is lost to heat—and not only in practice, but in theory as well. Speaking of theory, this is the meaning of the second law of thermodynamics, which says in essence that no actual, physical engine can convert heat energy completely into work. Some heat is always lost. Thus we can summarize theses two laws as follows: 1) You can't win, and 2) you can't even break even.

Although these losses cannot be eliminated, they can be minimized in two ways. The first is to improve the efficiency of the conversion device. Efficiency is the amount of energy gotten out after the conversion, divided by the energy put in. We find, for example, that in 1970 the gross consumption of energy in the United States was 64.6×10^{15} BTUs. Of this, 32.8×10^{15} was turned into useful work and 31.8×10^{15} went off as waste heat, giving an over-all efficiency of 51 percent.

The maximum efficiency of a single system is set by the difference between the input and output temperatures. The larger the difference, the greater the efficiency. More specifically, the hotter the vapors are made to burn in an automobile engine, or the hotter the steam fed to a steam engine, and the cooler the gases of discharge, the more efficient the engine can be made.

The efficiency of steam engines has been raised from less than 1 percent in early models to about 40 percent in the large modern steam turbines, mainly by going to higher inlet temperatures (raising the temperature of the steam entering the turbines). But we are reaching a limit; modern steam units run at 1100° F. and 3500 pounds per square inch of pressure inside the boilers. Indeed, we are reaching not only a practical but a true, theoretical limit. Even if, over the next few decades, we manage to develop materials that permit still further increases in input temperatures, it is not expected that efficiencies can increase to any more than about 50 percent.

Now, a 40 percent efficiency means that 60 percent of the original energy contained in the fuel has gone off in waste heat. Present-day nuclear plants do even worse in this respect, being only about 30 percent efficient.

Attempts are being made to improve these figures. In fossil fuel plants, more effective ways of burning the fuel are possible. The so-called fluidized bed process, developed in Europe, is one approach. Here the coal, instead of being powdered and blown into the furnace, is sent in on a moving grate, with air circulated through it from below. This process, though more complex than the conventional one, is not only more efficient but also produces less pollution.

The second way to minimize the heat losses mentioned above is to eliminate one or more of the conversion steps shown in the diagram on page 32. Today the great bulk of our electricity is still produced much as it was at the turn of the century. A source of heat is used to turn an engine

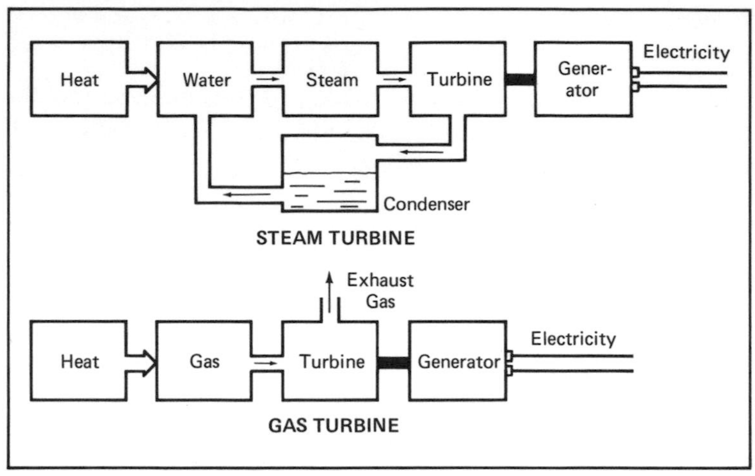

Basic heat engines.

(or turbine). As shown here, water is converted to steam which turns a turbine. The turbine (basically a windmill in a housing) is connected to a generator that produces the electricity. After most of its energy is used up, the steam is channeled to a condenser where it is turned back into water, and the cycle begins again.

It is possible to eliminate the steam cycle by use of a gas turbine. Though basically similar to the steam device, it uses the expanding hot gases directly, and so eliminates the need for any conversion step.

In both systems, however, the turbine is connected to conductors in the generator that are being turned through a magnetic field, and current is thereby generated in the conductors. And, in both systems, the method is rather cumbersome. Not only are efficiencies low, but rotating machinery tends to be noisy, prone to wear, and, often difficult to service and maintain.

Could the heat be turned directly into electricity without

going through the cumbersome steam/turbo-generator cycle? Could the energy of moving particles in a nuclear reactor somehow be converted directly into electricity? We turn now to the very important idea of direct energy conversion.

Direct Conversion

Electricity is, basically, a movement of charged particles, generally electrons. These are loosely bound in certain metals, such as copper, silver, and aluminum, and can be made to move easily by application of a voltage, which is why these substances are called conductors.

The temperature of a substance, on the other hand, is a measure of the average kinetic energy (energy in motion) of the individual molecules or atoms of which it is composed. Thus, the distribution and motion of electrons can also be influenced by temperature.

When two wires made of different metals are joined at their ends to make a circuit (as shown here), and the junctions are kept at different temperatures, a weak electric current flows in the circuit.

Although this effect was discovered in 1821–22 by T. J. Seebeck, the need and the means to put it to work had to await such applications as the space program, and such

Thermoelectric (Seebeck) effect.

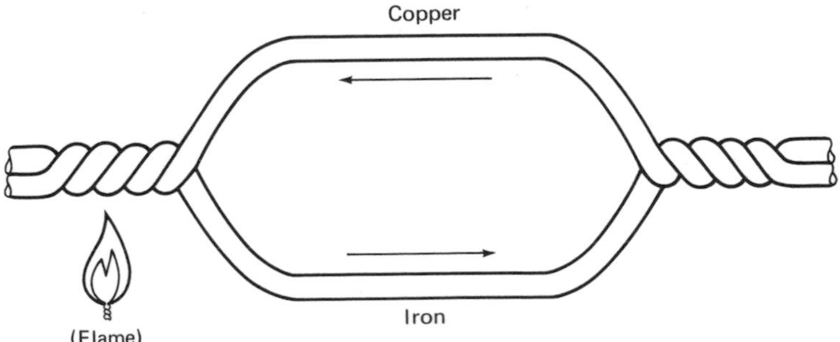

developments as semiconductors, which are fair conductors of electricity but poor conductors of heat. The phenomenon is now called the Seebeck or thermoelectric effect (thermo = heat) and is the basis for a promising means of converting heat directly into electricity, at least for some specialized applications such as space satellites. In several of the well-known series of SNAP power supplies (*s*ystems for *n*uclear *a*uxiliary *p*ower), nuclear fission was used to provide heat which was turned directly into electricity by a thermoelectric generator.

The thermoelectric effect occurs in the temperature range of about 0° F. to 1000° F. When a metal is heated to a higher temperature another phenomenon comes into play. At about 1800° F. electrons begin to "boil" out of the metal. This phenomenon, called the Edison effect, leads to another type of energy-conversion device, the thermionic converter.

A SNAP thermoelectric converter module. Heat brought in by a molten sodium/potassium mixture is converted to electricity in the thermoelectric element. Waste heat is radiated into space.

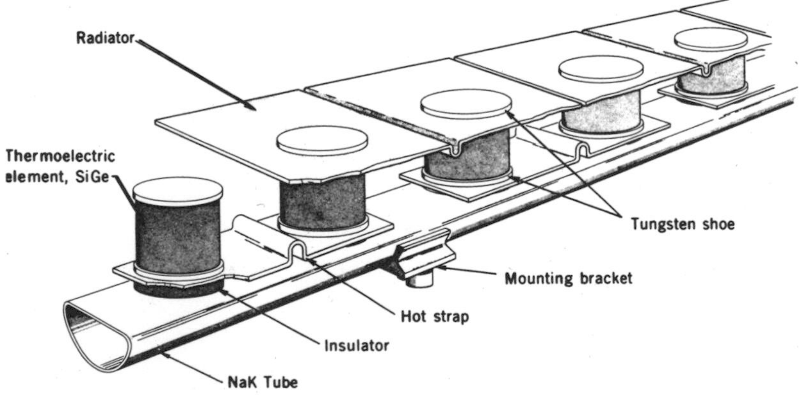

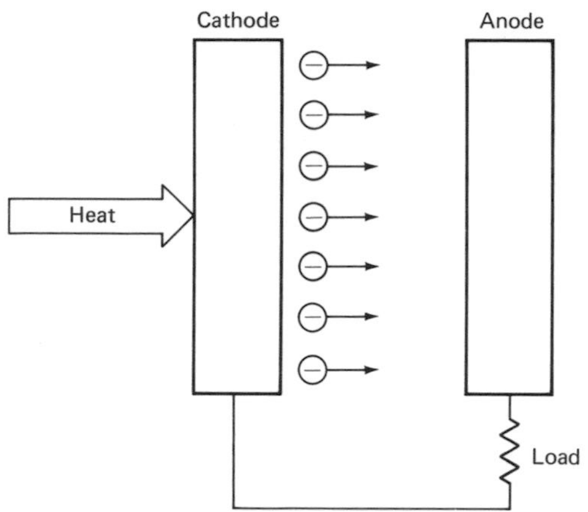

Principle of a thermionic converter.

(Any charged particle, positive or negative, is an ion. Ions driven out of a metal by heat are called thermions.) The principle of the thermionic converter is very simple and is shown in the accompanying diagram. Basically it is an electron tube, except that the electron emitter (cathode) is heated by combustion or nuclear processes, and instead of performing amplification or some other such electronic process, it converts heat to electricity. The anode collects the electrons.

There are difficult problems connected with the operation of materials at multi-thousand-degree temperatures, and in fission-heated devices the problems are compounded by radiation damage. Nevertheless, both thermoelectricity and thermionics show promise—they are inherently quiet, rugged, reliable, and easier to maintain than conventional electric plants.

They are not likely to be used as central station power plants because they actually have lower efficiency than conventional converters. But the higher efficiencies of the steam-electric plants are only found in the very large plants. As a result, in the direct conversion processes, which have already found use in the high-technology areas of space and military applications, may one day provide an important method of electricity generation in rural areas, where only small plants are needed. There are also tremendous new markets in the area of "portable" power. Of the two methods thermionics has potentially higher efficiency, perhaps 25 to 35 percent, due to its higher operating temperature.

MHD

We come, finally, to a third temperature-related effect that is much more likely to find widespread use in the generation of electricity. With the recent development, mainly through space work, of high-temperature materials and closer study of the motions and interactions of hot gases, a whole new approach is taking shape.

When a gas is heated to a high enough temperature, some of the electrons are completely stripped off the atoms. The resulting gas (or plasma), while still electrically neutral, now contains both positive and negative charges (ions). The negative charges consist of the electrons, and any otherwise neutral atoms to which they might attach themselves; the positive ions are those atoms that have lost one or more electrons. If these charged particles can be made to move under some form of control, we have an electric current.

The term that has come into use to describe the motion of charged particles in this context is magnetohydrodynamics. While a jawbreaker when looked at as a word unit, like many problems, it breaks down easily: magneto for magnetism, hydro for fluid (a gas is a fluid), and dynamics for the study of the motions and interactions of the particles. To save time and energy, engineers generally refer to the study as MHD, and we shall too.

In an MHD generator a stream of hot, ionized gas, such

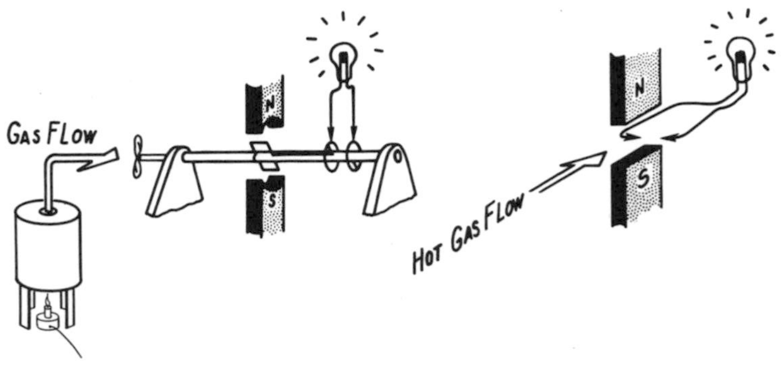

In a conventional generator, electricity is derived by rotating a solid coil of wire or armature, through a magnetic field. The MHD generator substitutes a high-temperature ionized gas for the armature, passes this gas through a magnetic field, and generates electricity.

as the gaseous product of a combustion process, is passed through a strong magnetic field (see illustration). The magnetic field causes the ions to veer to the left or right, depending on whether they are positive or negative ions; they are collected on electrodes and constitute an electric current, thus dispensing with the usual turbine/generator combination.

But all of this is more easily described than done. For one thing, high temperatures are required, on the order of 5000° F., which cause difficult materials problems. And even at these temperatures the gases have unacceptably low conductivity (too few ions) and must be "seeded" with a small amount of cesium or potassium, which ionize more readily. But these elements are too expensive to be disposed of after use and so must be recovered.

The desirability of MHD stems from several factors. First, this is a direct conversion system, which eliminates the need for such intermediate steps as turbines and generators. Thanks to a lack of moving parts coming in contact with hot fluids (which tend to cause corrosion and other problems at a far faster rate than cool ones do), the whole system is simplified and higher temperatures can be used, which leads to higher system efficiencies. There is also less pollution produced in MHD.

Useful Combinations

The various conversion devices we have discussed so far, including the conventional steam-turbine types, all operate in different temperature ranges. This suggests the possibility of combining them. Thus the hot gases that have already been used in an MHD generator can then be used to drive a thermionic system, and then again in a thermoelectric device.

Or, and this is the most likely first type of application, the MHD system can be used as a "topping" device for a conventional plant, meaning that its hot, gaseous exhaust can be used to drive a gas turbine. This double, or even triple, use of the hot gases is an excellent way of improving efficiency. It is expected that efficiencies as high as 60 or even 70 percent can be obtained in this way.

A dozen years ago MHD appeared to be the wave of the future and a number of research projects were initiated in the United States. For various reasons interest began to wane here, while it picked up in foreign countries. Thus it turns out that Soviets early in 1972 completed the world's first commercial MHD generator.* They are reported to be

* There are reports of severe engineering problems, however.

planning a $300 million combined MHD/steam-power plant, probably a more advanced version of the one just built.

The United States has no such grandiose plans. The problem has nothing to do with getting such a system to work. A working model was demonstrated in the United States back in 1959. The problem, aside from the high temperatures, corrosion, and so on, is that MHD is most efficient in large units, and so pilot plants and development projects are not an attractive investment for private companies.

With the increasing awareness of energy problems and shortages, it has been realized that more should be done in this area. Nevertheless, the closest we have come to a large project is a $2.4-million contract that Avco Corporation,

This experimental MHD generator produced 23.6 million watts of usable DC power.

a pioneer in the field, signed with the U.S. Office of Coal Research to develop an MHD system. MHD has also been pointed out as a specific area of interest in the President's budget request for the 1972–73 fiscal year.

The high temperature of a rocket exhaust makes it a logical place for possible application of MHD, and we may see its use for generation of electricity on large spacecraft. Because present types of nuclear plants do not produce very high temperatures, they are not good with MHD. We shall see later, however, that a future step in nuclear development, fusion power, may be an ideal complement to MHD.

There is also interest in application of the relatively recently developed liquid conductors, such as liquid sodium, to the MHD process. Instead of passing the hot combustion gases through the apparatus and out of the stack, some form of closed system would be used in which a high-velocity stream of liquid metal or perhaps another gas continuously circulates in a closed set of coils. In the Soviet device liquid potassium is used as the working fluid.

An allied direct-conversion process, electrogasdynamics, is also being considered. The EGD converter uses electric rather than magnetic fields to obtain a voltage. Unfortunately, although the voltage produced is high, the current is so small that power output is negligible. Because it is quite similar to MHD, and because there is less interest in it at this time, we will not go into it here. It is worth mentioning, however, that a high gas temperature is not a basic requirement in EGD. We may, therefore, be hearing more about it in the future.

4

Fossil Fuels

A SINGLE 380,000-kw electrical generating unit at Northport, Long Island, requires a total of 1 million gallons of oil per day; a large 1.2 million-kw (or 1,200-megawatt) coal-fired plant uses fuel at the rate of 8000 tons of coal—about 100 railroad cars full—every day. And there are some 3000 power plants of all kinds and sizes in the United States alone.

Yet we have seen that, at the moment, anyway, electric utilities use only one-fourth of all our primary energy resources. We use oil, natural gas, and even coal to heat our homes, offices, and factories, gasoline and diesel fuels to run our cars, trucks, and buses, and so on.

Since the turn of the century man has been on an energy spree. The result is that most of the consumption of fossil fuels in the history of man has occurred in the last quarter-century—and already there is talk of shortages!

The rate of production and use has varied for the various fuels. Coal use has remained relatively steady for the last 50 or 60 years, while oil, first, and then natural gas have gradually taken on more and more importance. Thus we find that the amount of coal produced and consumed since 1940 is roughly equal to the total consumption up to that time, while with oil the same holds true for the last decade!

The actuality and threat of shortages show up in various ways. There are, of course, the major blackouts experienced in the Northeast in 1965 and 1967, as well as the brownouts (drops in voltage) that took place quite often during peak-use times in recent summers. But perhaps even more significant is a fact like the following: In 1945 only 25 new field wildcat wells were required to make a significant discovery of oil; by the 1960's this had increased to 65. Also the average amount of oil found per foot of exploratory drilling has decreased markedly. In the decades around the turn of the century, the figure was 194 barrels per foot. In the 1920s it declined somewhat to 167. And in the thirties, thanks to some large finds, the figure rose to a high of 276. It has now dropped down to 35!

Total oil production 1859-1972 vs. reserves at end of 1972.

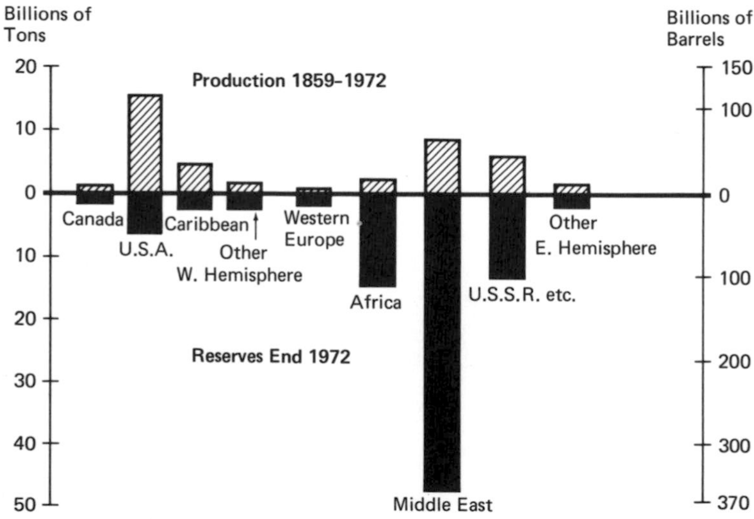

It seems reasonable, then, to ask, "How soon will it all be used up?" And indeed predictions are sometimes made; for example: U.S. oil reserves should last about 20 to 30 years, natural gas perhaps 35, and coal several hundred years.*

But this is too simplified a picture. The real answer is that these resources will never really be used up. What actually happens is that the most easily available and highest grade (most concentrated) resources are used up first. It must be kept in mind that it takes energy to discover, take out, and process these resources. Suppose the total amount of energy used to produce a barrel of oil or a ton of coal was the same or even more than that contained within the fuel. Clearly, exploration and production in this fuel would have stopped, for economic reasons, long before this point was reached.

Reserves

There are what might be called three levels of fuel reserves. A *proved reserve* is one whose location and approximate size has been more or less determined by drilling. It's as close as we can come to knowing how much fuel is there, short of actually digging up the top few miles of the earth's surface. The term *reserve* is a broader one, and refers to the resources thought to be recoverable under present, or at least given, demand, economics, and technology. The third term used is *resource,* meaning the total amount of a certain fuel which is thought to exist.

Clearly much depends both on rate of consumption and rate of discovery. If rate of consumption exceeds rate of discovery, proved reserves will decrease. And vice versa. The

* Harry Perry, "The Energy Crisis," in the *1972 Britannica Yearbook of Science and the Future,* p. 134.

history of prediction of resource depletion is littered with the corpses of incorrect forecasts based on the idea that consumption would increase while reserves would not. This is not necessarily the case at all.

First of all, new types of fuel or energy resources seem, at least historically, to have come in and taken over existing ones, as coal took over from wood. Unexpectedly rapid development of one or more of the energy sources we cover in this book could lead to decreased consumption of fossil fuels. Imports also complicate the picture. Thus while the United States is rich in coal, fairly well off in oil (for a while), and not so well off in gas, it can, for a price, import oil and gas from abroad, which it has been doing. While there are dangers and political implications (which we go into in more detail later) with imports, the point is that fairly large reserves of these fuels do exist in the world.

Another complicating factor is economics. A deposit of coal that is uneconomical to mine at a given price becomes worth going after at a higher price. The present concentration on a clean environment has put emphasis on clean fuels, which is an important reason for our present shortages. Should technology find ways to clean up dirty fuel economically, then presently unusuable fuel becomes usable, though more expensive.

So reserves can increase, and they do actually fluctuate somewhat. But the American Gas Association reported in 1972 that proven gas reserves fell in 1971 for the third time in four years (because consumption exceeded new discoveries).

It has been charged that the gas companies (and the oil companies, too) have deliberately held back on production to drive up prices. Spokesmen for these industries of course deny this, and indicate that under present conditions prices are almost sure to rise in any case.

Unfortunately we seem to be using up our more limited

fuels first. Oil and natural gas, which represent only about 9 percent of the total estimated United States supply of fuels, are being used at more than twice the rate of coal, which makes up 73 percent of our fuel resources.

Yet there has even been a shortage of coal. But it is a shortage of coal on hand. Again, the reasons are manifold. There was, for instance, a too-optimistic prediction of the rate at which nuclear power would take over from the fossil fuels in the generation of electricity. One result was an increase in the export of coal. Another was a drop in interest in coal production—fewer mines being opened, a drop in railroad car production, and so on. Combined with this was an unexpected consistency of demand.

Depletion

Nevertheless, in broad outline the fact remains that there is a real threat not only of shortage, but of actual, virtual depletion of the earth's fossil fuel resources within the next century or two. One can ignore this threat and say that future generations will probably have entirely different ways of producing their energy, and that it is ridiculous to worry about such a thing now. Others, more "future oriented," look at the matter differently, saying that we do have a responsibility to our descendants.

There is clearly no right or wrong. But for those who do think about such things, and who feel that what we do today has a strong influence on what happens tomorrow, it is important to have some logical way to consider the matter. Thus if present and projected rates of consumption are combined with some reasonable (it is hoped) estimate of total recoverable reserves, it is possible to calculate the probable length of time during which each fuel will be in use.

Estimating total supply is an extremely difficult task, and it is reasonable to expect different estimators to come up with different figures. Dr. M. King Hubbert, research geophysicist with the U.S. Geological Survey, makes an interesting point, however. He says that in terms of how long the supplies will last, the differences do not matter very much! Two very different estimates, for instance, have been given for oil. One is a figure provided by Hubbert (1,350 × 10^9 barrels); the other, by W. P. Ryman of Standard Oil, is almost twice as high (2,100 × 10^9 barrels). For the smaller figure a peak production rate of about 25 billion barrels per year is assumed for the peak year 1990. And the middle 80 percent of total (cumulative) production covers only the 58-year period from 1961 to 2032 (see diagram). For the second figure, which is almost twice the first, and which in-

Cycles of world oil production for two different estimates of total supply.

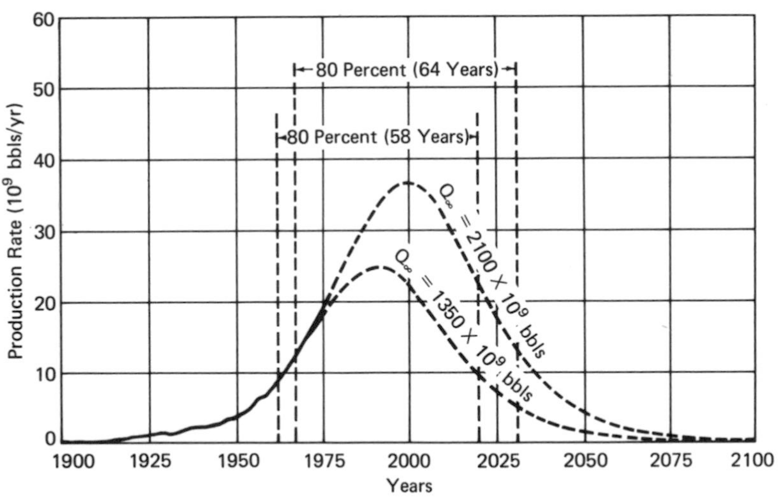

cludes a peak use of 37 billion barrels per year, the peak year shifts only ten years!

Consider then the reportedly huge discoveries of oil and gas made in Alaska in recent years. Even if they should increase the total U.S. supply by 50 percent, the over-all effect would be to postpone peak and trailing-off years by a decade or less!

Figures for coal, oil, and natural gas are given in the accompanying chart.

Fossil fuel use.

Fuel	*Total Recoverable Supply*		*Year of Peak Production*	*Year When 90% of Total Supply Will Be Used Up*
	Actual Estimates	*Converted to BTUs*		
NATURAL GAS				
U.S. (excluding Alaska)	$1{,}075 \times 10^{12}$ cu. ft.	$1{,}110 \times 10^{15}$	1980	2015
CRUDE OIL*				
World	$2{,}100 \times 10^{9}$ bbl.	$12{,}180 \times 10^{15}$	1990	2020
U.S. (excluding Alaska)	200×10^{9} bbl.	$1{,}160 \times 10^{15}$	1966	2000
COAL				
World	8.4×10^{12} tons	$210{,}000 \times 10^{15}$	2125	2380
U.S.	1.6×10^{12} tons	$40{,}960 \times 10^{15}$	2220	2440

* From estimates by W. P. Ryman of Standard Oil Company of New Jersey

** From data compiled by Paul Averitt of the U.S. Geological Survey

Alternate Sources

A large potential supply of oil lies chemically locked up in the earth. This was indicated in something that happened to a mining magnate in the 1880's. Matt Callaghan ordered a new house built, with a beautiful fireplace of polished marlstone. During the housewarming ceremonies, it being a cool night, he lit a roaring fire in the fireplace. Unfortunately no one told him, or no one knew, that marlstone is representative of a type of rock called oil shale, which contains a significant percentage of oil locked up chemically within it. The combustible material remains there, as in wood, until a high enough temperature ignites it. Callaghan's housewarming was a rip-roaring one indeed; the whole house burned down.

Although there are great reserves of oil shale, particularly in the United States, the shale contains only about 30 to 65 gallons of oil per ton, which means that an enormous amount of ore must be mined for each unit of oil obtained. Indeed, a greater volume of material is returned to the land after processing than was taken out! (This is because it was originally in more compact form.) The oil also contains undesirable materials which pose special problems in refining.

Methods of extracting the oil without having to dig up the shale are under consideration. It has been proposed that holes be dug down through the shale, and that fires be set at the bottom in such a way that the heat converts ("retorts") the material to a liquid that can be pumped out. Controlled nuclear blasts are another possibility. (Nuclear blasts have already been tried experimentally in an attempt to increase the production of natural gas.)

Another potential petroleum source is tar sands, which contain free oil but in such a viscous state that it cannot be removed as a liquid. A commercial plant producing some 50,000 barrels of oil a day has been in operation in Canada since 1967.

Tar sands may contain some 300 billion barrels of oil, and oil shale perhaps 190 billion barrels. Comparison with total requirements indicates that while these are large numbers, tar sands and oil shale are not likely to make any major dent in the supply situation, even assuming that the waste and other problems can be solved. It may be that these resources will prove more important as a source of raw materials for the chemical and manufacturing industries than as a source of fuels. Nevertheless, a number of major oil companies have purchased or leased large acreages of oil shale in Colorado, which certainly indicates at least a long-term interest in its possibilities.

A potentially large supply of natural gas may also lie over large areas of the ocean bottom. Columbia University reports work with an ice-like substance, gas hydrate, which forms at temperatures well above normal freezing when certain gases (including fuel gases) and water are combined under elevated pressures. A recent report by Soviet scientists has indicated an interest in large deposits of gas hydrate located in Siberia.

The Fossil Fuel Epoch

Summarizing the role of fossil fuels in human history, Dr. Hubbert considers a time period of 5000 years in the past to

Epoch of fossil fuel use.

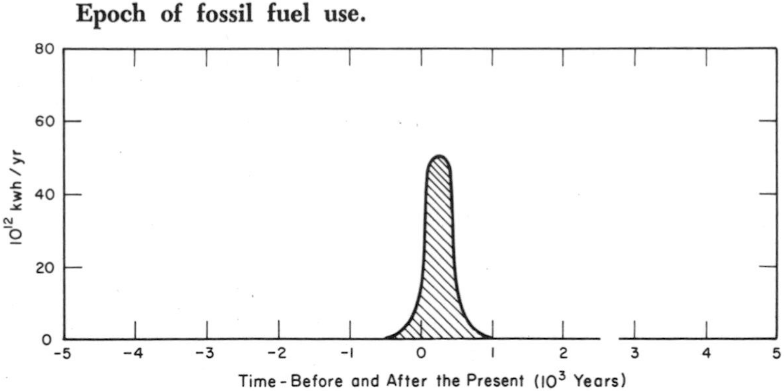

5000 years in the future—"a period well within the prospective span of human history. On such a time scale, it is seen that the epoch of the fossil fuels can only be a transitory and ephemeral event. . . ." * See diagram below.

It seems fairly certain, however, that fossil fuels will continue to dominate the fuel picture through the end of the century. After that the picture becomes far more cloudy. We will almost surely begin to see oil and natural gas begin to phase out. Although reserves of coal appear to be adequate for several centuries, at least in the United States, it is by no means certain that this fuel can continue to be widely used, mainly because of pollution, but also because digging for the less available supplies will entail increased mining problems. Considerably more research is needed in the coal area.

Nuclear energy is the presently available alternative, or supplement. But it has, as we shall see, problems of its own.

* *Resources and Man,* National Academy of Sciences–National Research Council, p. 205.

5

Fission

THE RISE OF NUCLEAR POWER has been nothing less than astonishing. The first physical demonstration that it could be done took place in 1942. The first commercial-size plant, at Shippingport, Pennsylvania, began generating 60,000 kilowatts of electricity in 1957, just 15 years later. The first truly large-scale plant—a 428,000-kilowatt installation at San Onofre, California—was licensed for construction in 1964 and went into operation in 1967. Today 1-million-kilowatt units are already fairly common.

This would be equivalent to compressing the 67 years that elapsed between the first Wright Brothers' flight and the Boeing 747 into 30 years.

Further, although nuclear power still generates only about 3 percent of our electricity, most forecasts maintain that that figure will climb rapidly—to perhaps 25 percent by 1985 and as high as 50 to 60 percent by the year 2000.

To understand what is so special about nuclear plants, we must of course have some understanding of what they are and how they work. This will also help us to look into the future of nuclear power.

Some Background Information

Today practically all the commercial energy used by man still comes from some form of combustion. Combustion is a chemical reaction, examples of which are fire, metabolism, and explosion. Cars run, for example, as a result of a timed, controlled series of explosions in the engine cylinders.

Around the turn of this century, after the discovery of radioactivity, the heating effect of radium emissions was measured and found to be millions of times the energy available in ordinary chemical reactions. Basically, the difference lies in the fact that radioactivity has to do with changes in the nucleus itself rather than in the atoms and molecules of the substances concerned.* And the forces involved—the so-called nuclear binding forces which keep the neutrons and protons together—are far more powerful than the electromagnetic forces involved in chemical changes.

Neutrons have no charge, but protons are charged positively, and like charges repel each other; therefore, nuclei are being acted upon by both repulsive and attractive forces. In most cases the result is stability. In large nuclei, however, the greater number of protons begins to exert enough repulsive force to create a kind of instability, which results in such things as radioactivity. Now, if we try to shoot another proton into a large nucleus, we have a hard time because the positive charge of the existing protons tends to repel the proton. But neutrons are not charged and can enter a nucleus with no trouble at all. The result, in such elements as uranium and plutonium, is a tendency of the nuclei to split, or fission, into two parts.

* This is why purists object to use of the term *atomic* energy and prefer *nuclear* energy.

This, in itself, would be interesting but not terribly useful. In addition to the fission fragments, however, two or three neutrons are also shot out. Thus not only do we have the kinetic (mechanical) energy of the fission fragments, but we have additional neutrons which, if handled properly, can keep the fission process going!

An important point to understand here is that a nuclear plant is really very similar to a fossil fuel plant, except that the source of heat is nuclear fission rather than chemical combustion. What happens is that as the fission fragments collide with surrounding matter, their kinetic energy is turned into heat. This heat is transferred to water, turning it to steam, which in turn is used to drive turbine generators.

Thus the essential parts of a reactor are:

1) A core in which the fission reaction takes place.

2) The fuel, which fissions, producing neutrons and heat energy.

3) Control elements, which control the rate of release of this energy.

4) A cooling fluid, which removes the heat generated.

5) The reactor vessel, which contains the whole process

Simplified fission process.

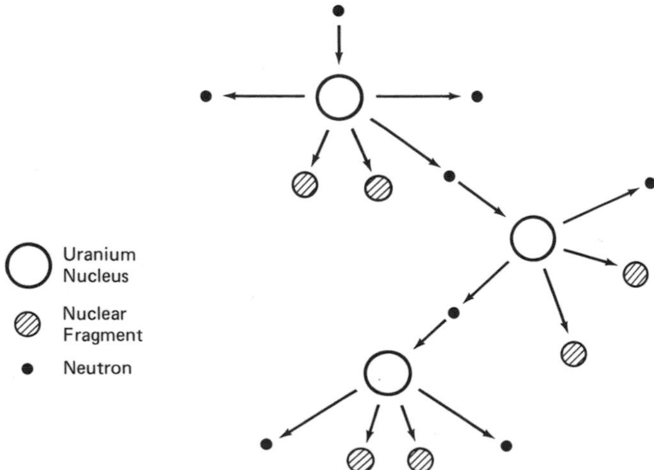

and protects the outside world from the heat and radioactivity being generated inside.

Reactor Types

Although a large number of reactor types have been designed and tested, utilizing various kinds of fuel and coolant

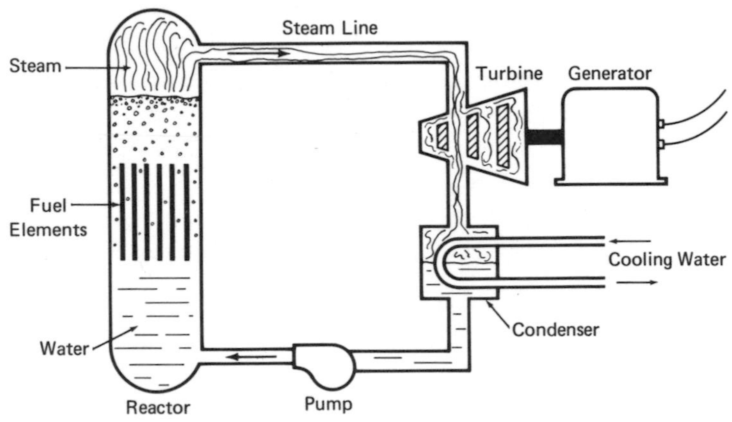

In a boiling water reactor (BWR), the reactor itself is the steam generator. The water and resulting steam are forced through a group of metal-clad, slightly enriched uranium dioxide fuel elements and into the steam line by a very powerful pump. The steam, which provides the mechanical energy to drive the turbine-generator, is heated to about 550° F. at a pressure of 1000 pounds per square inch. This is about equal to the pressure half a mile down in the sea. By comparison, steam in a tea kettle at atmospheric pressure is produced at 212° F.; steam at this temperature has too low an energy value for use in a turbine. The higher pressure in the reactor raises the boiling point of the water, so that the temperature can be kept at a higher point.

Nevertheless, the water is allowed to boil in the reactor, hence the name. This eliminates the need for a separate loop, as in the pressurized water reactor, shown on the next page, but the physical size of the core must be larger. An 800,000-kw unit, for instance, needs a reactor vessel about 70 feet high by 20 feet in diameter.

systems, almost all the commercial ones have used uranium as the fuel and water as the coolant. The two basic types are shown and described in the accompanying diagrams.

Strangely enough, these two types, the PWR and BWR, are the least efficient of the various types investigated. They are, however, basically the simplest to design and build, and their technology was furthest along when commercial plants

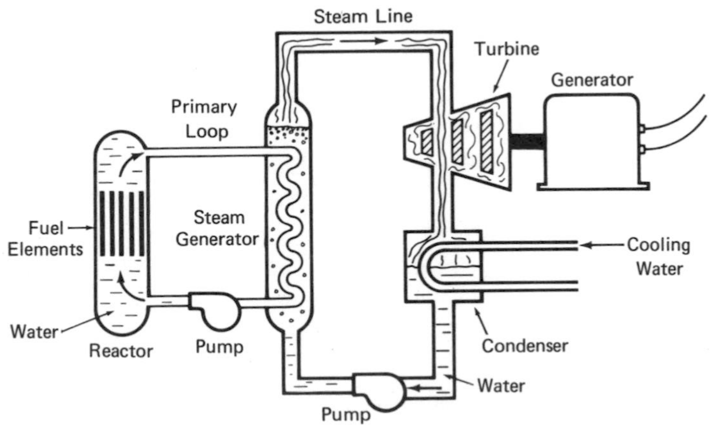

The pressure in a pressurized water reactor (PWR) is kept at about 2000 pounds per square inch, which permits the water to be heated to 600° F. Thus the water is not permitted to boil in the reactor, as in the BWR, but goes instead to the steam generator, where the heat is transferred to water in a separate loop, converting it to steam at about 500° F. It is this steam that drives the turbine-generator, while the water in the primary loop goes back to the reactor to be heated once again. The volume of the pressure vessel for a PWR is about one-third that of the equivalent BWR, although the additional equipment needed more or less makes up for this. Though the PWR pressure vessel is smaller than that of the BWR, its higher pressure requires steel walls about a foot thick. Such a vessel may weigh a million pounds empty, and more than a million and a half pounds with the water and core in place.

began to be built. Also, water has been generally cheap and abundant, and uranium has been available from bomb plants. Other types of reactors, which we'll get to in a moment, have begun to enter the field.

There are two major problems inherent in the use of ordinary water as the coolant. One is that, for both technical and economic reasons, it is difficult to obtain high-pressure, high-temperature steam in nuclear plants. Water-cooled nuclear plants thus produce what is called low-quality steam; the result is that they are not as efficient in converting heat to electricity as are present-day fossil fuel plants. A water-cooled nuclear plant discharges about 40 to 50 percent more waste heat per kilowatt hour of electricity than a modern fossil fuel plant, and thus requires proportionately more cooling water. (Actually, all conventional, water-cooled plants require great amounts of cooling water, which is rapidly becoming a problem; it is becoming increasingly difficult to find suitable sites for power plants.)

The second limitation of water-cooled reactors is that they are not efficient in their use of nuclear fuel. Now this may sound strange, especially in the light of the often heard comparison that one pound of uranium (the size of a golf ball) can produce as much heat as 3 million pounds of coal.

The important word here is "can." Naturally-occurring uranium is found in three forms. They are the isotopes uranium-234, -235, and -238, with abundances of 0.006, 0.711, and 99.283 percent respectively.* All three are chemically similar, though they have different numbers of neutrons in their nuclei; the number 234 refers to the total number of neutrons and protons combined. Of the three, unfortunately, only the 235 form is fissionable. Current American designs

* An isotope is a variety of an element that has the same number of protons as the other forms, but a different number of neutrons.

have taken advantage of the existence of uranium-processing plants from our earlier bomb efforts. These have been used to produce quantities of slightly enriched uranium, in which the percentage of fissionable material has been increased to 2 or 3 percent.

Uranium-238 is not fissionable, but is fertile, which means it can be converted to a fissionable material, plutonium-239. Similarly, thorium, which is fertile, can be converted to fissionable uranium-233 (which is not found in nature).

In other words, only about 1 uranium atom in 140 in the natural state is fissionable. Although PWRs and BWRs do convert some uranium-238 to plutonium, by far most of their power must come from the 235 form. The reason is that the water used in the reactor for cooling has a strong tendency to absorb neutrons. Thus while there are enough energetic neutrons left to keep a chain reaction going, there are not enough left to produce a significant amount of conversion. Indeed, these plants consume about twice as much fissionable material as they produce.

As long as fossil fuel costs are high and uranium costs comparatively low, and as long as uranium fuel is relatively abundant, this type of cycle still makes economic sense, especially in the larger, 1-million kilowatt plants now being built.*

But supplies of uranium-235 are not abundant, and if we continued to depend on PWRs and BWRs, it would not be long before shortages of easily available ore developed, and prices went up. Dr. Hubbert feels that an acute shortage of low-cost ores would develop before the end of the century. Indeed, if the nuclear approach were to depend en-

* Because building costs are high, while fuel and fuel-transportation costs are low, the larger the nuclear plant, the cheaper the per-kilowatt cost. We can expect to see 2-million-kilowatt plants being built before the end of the 1970's.

tirely on uranium-235, its day in the sun would be even shorter than that of oil and natural gas.

Nuclear systems have been designed to take advantage of the conversion possibilities of uranium-238 or thorium. These are called advanced converters, and produce about as much fuel as they consume. At this time the one that is best known and most highly developed is the high-temperature gas-cooled reactor (HTGR), of which several have been built in recent years. The HTGR can also operate on uranium-233. Thus it adds the abundant element thorium to our nuclear fuel reserves.

The HTGR offers other advantages. Because it uses gas as the coolant, helium in this case, and graphite rods for containing the fuel (rather than a metal alloy), it can operate at higher temperatures, thus making it a more efficient machine—39 percent rather than 30 to 32 percent. Among other things this means, as we have already mentioned, less thermal (heat) discharge. Because the HTGR is gas-cooled, it also becomes possible to use the gas directly to turn gas turbines, thus eliminating the cumbersome steam cycle. All in all, the HTGR is expected to have fewer environmental problems than the water-cooled machines.

A demonstration 40,000-kilowatt HTGR plant has been in operation since 1967 at Peach Bottom, Pennsylvania. A larger (330,000 kilowatts) and more advanced unit is being completed at Fort St. Vrain, Colorado; and in 1971, four units of this type were ordered, the largest of which were a pair of 1,160,000-kilowatt units for Philadelphia Electric Company.

Breeders

Carrying the idea of preserving our nuclear fuel supply a large step further is the so-called breeder reactor. A breeder is a reactor that creates more than one atom of fissionable material for each atom of fuel consumed. (This is *not* a kind of perpetual motion, but merely a more efficient use of the energy contained in the fuel.) Recall that for every atom of fissionable uranium-235 there are 140 of fertile uranium-238, and that thorium can be used as well.

The point of the breeder is that it is designed to utilize the neutrons that, in conventional reactors, are now absorbed or otherwise lost. By the same token there are, as in non-breeders, a number of approaches to the desired end. The one with the most promise at this time is the fast-neutron type, or, more simply, the fast breeder. The important point to understand is that all the neutrons emitted during the fission process

An engineering mechanics technician takes precise measurements on the mock core of a fast-breeder nuclear reactor being developed at the Hanford Engineering Development Laboratory, Richland, Washington. Westinghouse was chosen to operate the lab for the U.S. Atomic Energy Commission. Fast-breeders are expected to be a widely used type of reactor for power generation in the 1980's.

are fast neutrons; they are traveling at 30 to 40 million miles per hour. As it turns out, however, the fission chain reaction in uranium-235 is not well served by fast neutrons, and so water is used not only for cooling and heat transfer, but also as a moderator—that is, to slow down the neutrons to "thermal" speeds, which are about 5000 miles per hour. The thermal neutrons tend to "ignore" (float past) the more abundant uranium-238 atoms and to concentrate on the 235 atoms. Thus PWRs and BWRs are sometimes called thermal reactors.

Obviously the term *fast breeder* means that the neutrons emitted in the fission process can be used without having to slow them down first, which means in turn that water cannot be used in the system.

Although other systems are being considered and even developed, the one that has taken a strong lead is the Liquid Metal Fast Breeder Reaction or LMFBR, and so we shall describe that one here.

A number of reasons have been advanced for this choice. Although liquid sodium is more expensive than water, it is still quite common (the sixth most common element). It is an excellent heat-transfer medium. And its naturally high boiling point (1616° F.) permits high-temperature, low-pressure operation.

But the main reason has been (and this is similar to the reason for the rapid growth of water-cooled thermal reactors) that considerably more experience had been developed in the laboratory here and abroad with liquid metal systems than with other coolants, such as gas, molten salt, and steam. This was a strong enough factor to overcome the fact that sodium has an extremely high chemical reactivity and will, for example, react violently if it comes in contact with water.

As a matter of fact, the very first nuclear reactor to ac-

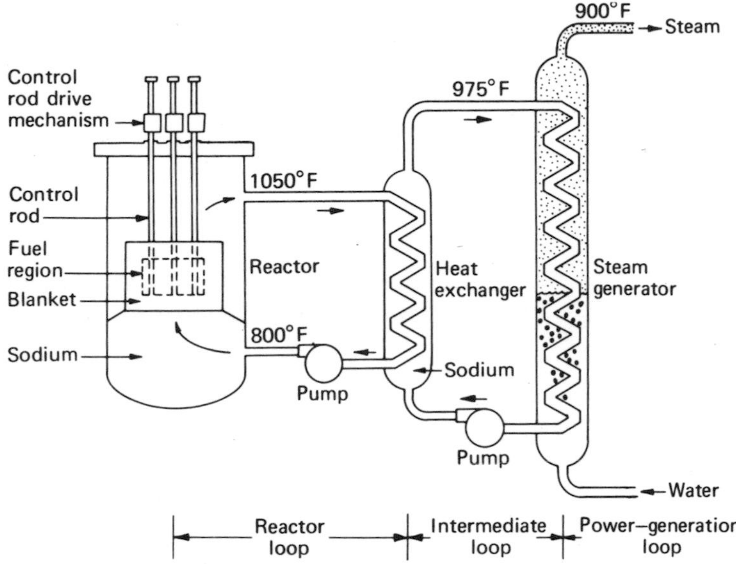

While breeder reactors are reactors in the same sense as the BWR and PWR, there are several important diffrences, which can be seen in the diagram.

The first difference is the extra, or intermediate, loop. This is necessary because sodium becomes radioactive in the primary system. The intermediate loop is a safety feature, ensuring that a leak in the primary system will not allow radioactive material to enter the power generation loop or into a chemical reaction with the water or steam contained therein.

The second difference is the "blanket" seen surrounding the core. As in conventional reactors, the core contains both fissionable and fertile materials; in this case it is about 20 percent plutonium-239 and 80 percent uranium-238. The blanket also contains uranium-238. The fuel initiates and sustains the chain reaction, producing excess neutrons in the process. These are absorbed in the blanket, converting it, atom by atom, into new fissionable material.

tually be connected to electric-generating equipment and to produce electricity was a liquid metal (sodium-potassium) fast breeder, the EBR-1 (Experimental Breeder Reactor).

As of mid-1972 there were 22 LMFBR projects in the industrial nations of the world: 8 are operable, 4 are being built, and 10 are planned. Most are experimental or relatively small demonstration reactors. But the Soviets have just completed the world's first large commercial fast breeder reactor, a 350,000-kilowatt plant (of which 150,000 will be for electricity, and 200,000 for desalination of water). They have announced plans for a 600,000-kilowatt plant, to be completed in 1976.

In the United States a breeder demonstrator plant, of 300,000- to 500,000-kilowatt size, is to be built in the mountains of Tennessee by Commonwealth Edison of Chicago and the Tennessee Valley Authority (an independent government agency). An extremely important aspect of this project is that the private utilities have pledged a contribution of $240 million toward its cost, indicating its commercial potential. Plans are to follow this first demonstrator with a second of equal size.

You will recall that in thermal reactors uranium-238 is converted by the absorption of neutrons to plutonium-239. In these reactors this is an unwanted product because thermal neutrons do not work well with plutonium. So another advantage of the LMFBR is that it will provide a market for the increasing amounts of plutonium being created in thermal reactors.

And, finally, the use of liquid metal means that the reactor can (indeed, must) operate at higher temperatures. It can thus be more efficient and will produce less thermal pollution.

Another measure of its efficiency is the "doubling time," the time required to produce as much additional fissionable material as was originally in the reactor. An efficient breeder will have a doubling time of 7 to 10 years. While this sounds like it should solve our fuel problems, we must remember, first, that there are still environmental problems to be solved, which we cover more fully later on, and, second, that our electric power requirements have been just about doubling every 10 years. So even with breeders we will barely be holding our own.

The fast breeders offer faster doubling times (higher breeding gains) than the thermal breeders. The two main thermal breeder concepts, using a thorium uranium-233 cycle, are the Molten Salt Breeder Reactor (MSBR) and the Light Water Breeder Reactor (LWBR). The MSBR is interesting in that it has no fuel elements as such; the fissionable substance is contained in a circulating loop of molten salts and only fissions when a particular geometry is attained, which of course is set up only in the core of the reactor. The LWBR is considered to be of interest because if it is developed the many PWRs we will see in coming years could perhaps be converted to breeders by merely changing the core.

Indeed, there are those who claim we have leaped too quickly onto the LMFBR bandwagon. Of the alternatives, probably the one with most support is the Gas-Cooled Fast Breeder Reactor (GCFR). This device can draw upon the rapidly developing technology of the HTGR (High-Temperature Gas-cooled Reactor), and as with the LWBR, conversion of gas-cooled thermal reactors to breeders would be relatively simple. It also eliminates the need for handling sodium. Whether the rapidly developing gas-cooled systems will be able to overcome the head start of the LMFBR remains to be seen.

While breeders hold out great promise, a lot of technical problems remain to be worked out for all the various types.

It is also not clear what the environmental problems may be. Although they are actually "cleaner" in that they emit virtually no radioactive materials into the atmosphere (as opposed to a small amount for water-cooled types), their fuel is plutonium, one of the most toxic materials known. Plutonium-239 has a half-life of 24,000 years, which means that half of any given amount is still radioactive after that period of time! By the year 2000 we may be producing and handling 100 *tons* of the material every year. Breeders are also more sophisticated and hard-to-manage machines.

Nevertheless, an Atomic Energy Commission prediction says that, by the end of the century, breeders will have a power capacity exceeding that of all present-day plants combined.

So, on the one hand, we have President Nixon's energy message of June 4, 1971, calling for what amounts to a national effort to develop an LMFBR by 1980; and on the other we have critics who say, "Slow down, you're goin' too fast."

Who is right? Only time will tell. In the meantime let us look more carefully at some of the problems faced by both fossil fuel and fission plants.

III. PROBLEMS WITH TODAY'S ENERGY

6

Air Pollution

IT IS PROBABLY FAIR to say that, at the present time, there is no over-all energy shortage in the United States. There is, however, a very obvious shortage of clean energy.

For example, the Los Angeles area and other sections of the Southwest are badly in need of new sources of power. Utah, Nevada, Arizona, and New Mexico have large deposits of coal. What could be more reasonable than to build large power plants in this area which would use this coal, and to ship the electricity to the more built-up areas?

The problem, however, is not supplying the electricity, which is indeed being done, but the fact that large quantities of pollutants are now being pumped into the once-pristine air of the Grand Canyon, Mesa Verde, and other one-of-a-kind natural treasures of our American Southwest. Already, sharp-eyed visitors are seeing a faint fogginess that has never been seen before. A study prepared by the Environmental Protection Agency warns that unless something is done (a number of other plants are under construction or planned in the area), the once perfectly clean air of the Southwest "will be increasingly and significantly degraded."

The situation in cities is far worse. Again the problem is more one of pollution than shortage. Virtually all city-dwellers—and visitors—have had the experience of stinging eyes, irritated throat, and various other problems brought on by toxic compounds in the air. In 1952, as a result of a peculiar atmospheric phenomenon, some 4000 people died in the London area from the air pollutants which were trapped there by an atmospheric phenomenon called an inversion. Similar situations, though not as bad, have been recorded in other cities at other times. The Los Angeles smog is too well known to discuss any further.

Damage to buildings and monuments is another problem, to which cities claiming a rich past are particularly vulnerable. A number of monuments in Italian and Greek cities have had to be closed, at least temporarily, because of danger of collapse due in large part to weakening from damage done by toxic substances in the air.

And all of this ignores other damage done to our water and land, which we'll get to later. The major causes of air pollution are given in the table.

Sources of Air Pollution
Millions of Tons Per Year (1969)

Source	*Sulfur Oxides*	*Particulates*	*Carbon Monoxide*	*Hydrocarbons*	*Nitrogen Oxides*	*Totals*
Transportation	1.1	0.8	111.5	19.8	11.2	144.4
Fuel Combustion in Stationary Sources	24.4	7.2	1.8	0.9	10.0	44.3
Industrial Processes	7.5	14.4	12.0	5.5	0.2	39.6
Solid Waste Disposal	0.2	1.4	7.9	2.0	0.4	11.9
Miscellaneous	0.2	11.4	18.2	9.2	2.0	41.0
Total	33.4	35.2	151.4	37.4	23.8	260.2

(Source: Office of Science and Technology)

Let us look into the air pollution problem more closely.

Transportation

In 1932, in the depths of the "great" American depression, Herbert Hoover ran for the office of the Presidency with the slogan, "A chicken in every pot and a car in every garage."

Today, to make any kind of impression, it would probably take a promise of a steak in every broiler and *two* cars in every garage. For cars have become the mainstay of our transportation system; there are already some 95 million of them on the road; gasoline for private cars represents between one-fifth and one-sixth of all the energy used in this country each year. (If all mobile sources, such as trucks and buses, are included, the figure goes up to about one-fourth.) Unfortunately, cars are extremely inefficient users of energy, requiring on the order of 5 or 6 times more fuel per passenger mile than trains and buses. Further, they are clogging our cities and suburbs to the point where movement often stops altogether.

At the moment, however, the biggest objection to the automobile is that its internal combustion (IC) engine, more than any other single factor, is polluting our air and making it practically unbreathable in at least some urban areas. In certain sections of Tokyo traffic policemen have to take oxygen after a 1- or 2-hour shift on duty. Some actually wear oxygen masks.

Among the approaches being taken to alleviate the situation are: use of natural gas as a fuel; electric cars; Wankel Stirling, stratified charge, Diesel and steam engines; and gas turbines. Hybrids are being considered too, such as a car that runs on an IC engine in the country, and on battery while in the city. An important advantage of a hybrid vehicle is that the engine part can be run throughout at essentially constant speed; this makes for greater efficiency in fuel use,

	ROTARY	V-8
Horsepower/RPM	185/5000	195/4800
Weight—lbs.	237	607
L x W x H—in.	18.0 x 22.1 x 21.5	29.5 x 28 x 31.5
Volume—cu. ft.	5	15
Number of parts	633	1029
Number of moving parts in power section and drive line	154	388

Comparison of rotary engine with a V-8 of comparable horsepower.

but also makes reduction of emissions an easier problem to solve than is presently the case.

The battery approach will be dealt with elsewhere in this book. We haven't the space to describe all of the others here, but information is widely available on all of them. The Wankel, or rotary, engine as it is sometimes called, seems to have the edge at this point. A Japanese car, the Mazda, has actually been put on the American market with such an engine.

The Wankel is in essence a simplified internal combustion engine. While it burns no cleaner than a standard IC engine, its exhaust gases are hotter, hence better suited for treatment. The engine is also smaller and so makes more space available for pollution-control devices. And, finally, it can operate on lead-free, low-octane gasoline.* The importance of the pollution factor in the auto field is indicated by the fact that the Wankel is even less efficient in fuel economy than IC engines, yet General Motors has already announced plans to offer the Wankel as an option in a year or two.

But the investment of the car companies in the IC engine is huge. They are not about to allow it to go down the drain without a fight. Pressed on the one side by more stringent emission requirements, and on the other by newer or at least different approaches to the problem, they have intensified their efforts to clean up the IC engine. As a result, and in spite of very recent statements by car company executives that a "clean" car could not be produced in the time allowed, things are looking better at the moment. General Motors, normally quite conservative and truthful in its claims, now says it may have the automotive air pollution problem licked—at least in demonstration cars. Whether carelessly made production cars will hold their own in this respect; whether the public will be willing to spend the extra money for the devices required (several hundred dollars per auto); whether they will stand for a 3 to 12 percent increase in fuel use, and a deterioration in over-all drivability*; and whether they will stand for the fact that the

* The main reason for the inclusion of lead in present-day gasolines is that this is the cheapest way to raise the octane or power rating; lead also serves as a high-temperature lubricant.

* Dr. A. J. Haagen-Smit, the scientist who figured out the causes of smog 20 years ago (mostly due to nitrogen oxides), says: "I don't want my car to stumble trying to enter a Los Angeles freeway at 70 miles per hour because of pollution controls. I am not afraid of dying from smog, but I am afraid of dying in an auto accident." (*Technology Review,* July/August 1972, p. 9.)

devices will probably have to be replaced or beefed up after each 25,000 to 50,000 miles of driving—all this remains to be seen. Also, say company executives, it will be necessary for the petroleum industry to make some changes in their gasoline, a costly and complex requirement. Certain of the pollution-control devices, for example, are quickly rendered ineffective if the fuel contains lead.

And, finally, we must realize that even if emissions are cut down in new cars, there will continue to be many old ones on the road and unless we restrict the continued proliferation of autos, we will produce the same effects due to increasing numbers.

For the more distant future, the electric auto still holds promise of pollution-free operation, assuming that pollution-free production of electricity can be accomplished and that some efficient means of storage or transfer of electricity can be found. Perhaps the vehicles can somehow pick up power from cables buried beneath the streets. Presently, the battery is the only method actually available.

Over-all efficiency of electric vehicles, including losses due to generation and transmission of electricity, may be no higher than with gasoline engines, and may even end up lower. Estimates vary. D. P. Grimmer and K. Luszczynski, physicists and members of the Committee for Environmental Information, maintain that "only about one-tenth of the raw fuel energy is delivered to the wheels of the gasoline car, as compared to one-fifth [for] the electric car." Others claim the reverse. In March of 1972, battery-powered electric buses went on trial in smog-filled Osaka, Japan, in an attempt to cut pollution. Perhaps some more definitive figures will come out of this.

Another problem is that the requirement for electricity would rise enormously. Grimmer and Luszczynski estimate that electrification of most of the motor vehicles in the

United States and half the railroad transport could be accomplished with an increase of only 25 percent in over-all generating capacity (since most of it would be done at night). But even this would severely tax our already-strained generating facilities. Other estimates are much higher.

If it could be done, then electrification of our transportation system would be a logical way to shift from fossil to nuclear fuel, should this be the route we decide to take. But even now electrical use for rapid transit in cities is far higher than most people think. The largest single user of electricity in New York City is the subway system; the Metropolitan Transit Authority takes fully 11 percent of Consolidated Edison's entire output.

It may be that some sort of combined public transport/individual car will be the ultimate, in which the driver operates the vehicle on its own power on local roads, but hooks on somehow to power and guidance on long-distance or heavily traveled roads.*

Power Production

A 40 percent efficient, 1 million-kw generating plant burns some 340 tons of coal an hour and produces in that time 940 tons of carbon dioxide, 17 tons of sulfur dioxide, 3.4 tons of nitrogen oxides, and 34 tons of ash. (These figures may vary somewhat depending on fuel type.) The ash can be used for building purposes, but this is not often done because other materials are actually cheaper and more convenient to use.

The major problems in power plants are sulfur dioxide, an irritating gas, and particulates (soot, tiny ash particles, etc.). According to government sources power plants are esti-

* See *Transportation in the World of the Future,* another book in the World of the Future series, for more on this.

mated to be responsible for some 50 percent of current sulfur dioxide emissions, 20 percent of particulates, and 20 percent of nitrogen oxides.

Before the turn of the century, the major concern was the emission of particulates. At the turn of the century, dense, even poisonous, smoke used to pour out of the chimneys of metal smelters and other industrial plants in the growing San Francisco region. Downwind areas were considered dangerous for humans, animals, and vegetation.

At that time pollution was, compared to now, more localized, more visible, and probably more severe in certain highly industrialized areas, such as Pittsburgh.

The invention of the electrostatic precipitator made it possible to clean up the more obvious of the emissions, thus permitting one to live in an urban area without getting full of soot and smoke particles.

Such devices are now capable of removing as much as 99 percent of particulates from the stack.* A remaining problem, however, is that what they are removing are the larger particles. The smaller ones, though making up a tiny fraction (by weight) of the emissions, are more dangerous in that they are more easily absorbed into the lungs. Thus, with the proliferation of power and industrial plants, even this major cleanup has not been sufficient to wipe out particulates as a problem.

Much more of a problem, however, is sulfur dioxide, due mainly to the burning in power plants of high-sulfur-containing fuels. A further complication is the fact that oxides of sulfur react with a number of particulates to form even more damaging and persistent compounds, such as sulfuric acid. Thus their hazard (to humans, animals, vegetation, and

* Some 20 million tons of this material are collected annually, of which about 1.3 million tons, or 6 percent, are used commercially.

building surfaces) is considerably greater than their amount would indicate. New York City in a recent study was found to have sulfur oxide levels 3 times higher than considered safe by Federal standards.

Although a number of methods are under investigation, no commercially practicable way of removing sulfur during or after combustion has been found. The worst offender is coal, though oil too may contain a high content (more than 1 percent). One electric utility in the Midwest is perched atop a large deposit of 2 percent sulfur coal which it cannot use because of recently enacted regulations. Indeed, most coal contains too high a content to meet present environmental standards. And 90 percent of the cleaner coal in the United States is found in the western section of the country, where it is needed least.

Even at a premium of $2 to $3 a ton in the East, producers may not be able to supply all the low-sulfur coal that would be needed, and the cost of electricity would rise considerably, putting fossil-fueled plants at a great disadvantage relative to nuclear power. (Most of the coal burned in this country is still high-sulfur coal.)

One helpful method of ameliorating the effects of this noxious substance is the use of tall stacks, which help disperse it and the other pollutants into the upper air layers. A *1,200-foot* stack, taller than the Empire State Building, has been built in West Virginia; but this is no real answer, especially with continuing growth in energy production.

Unless a practicable way is found to curb these emissions, we may eventually come to the point where it will not be possible to put up fossil-fueled generating plants anywhere, let alone use the immense coal resources in this country and elsewhere. Such plants are even now generally barred from badly polluted urban areas.

One hopeful approach is to remove the sulfur from the oil and coal* before burning it!

As for cleaning the heavy oil usually used in power plants (residual oil), the best present method seems to be to do it at the refinery. Oil can be desulfurized right now, but at an additional cost of 10 to 15 percent of production cost.

Regarding coal, the most hopeful approach is the liquefaction or, most likely, gasification of the solid fuel. That is, the coal is processed in such a way that most of its combustible material is separated and formed into a burnable gas (or liquid). What is involved is a relatively simple chemical process, the addition of hydrogen to change carbon to CH_4, which is the principal constituent of natural gas. But a simple chemical process is not necessarily an efficient or economical † one.

The importance of gasification is indicated by the fact that a good-sized effort is underway, with both government and industry resources devoted to it. Out of 7 new gasification projects begun or announced in 1971, 3 will use coal as the raw material, 2 will use natural gas liquids, and 2 will use naphtha. The last 4 are to be of commercial size and have established technology but are considered stopgap measures because the raw materials are petroleum products and are themselves in relatively short supply. Oil shale is another possible source.

Although the product may have the same makeup as natural gas, it is sometimes called syngas (synthetic gas) when produced in this way. One such coal-gasification plant, which has reached the demonstration stage, is shown in the photo.

* Burning of natural gas produces little in the way of sulfur and particulates, but does produce nitrogen oxides.

† In Europe, where natural gas costs more than in the United States, coal gasification has been going on for years. But the process used is too expensive for use here.

Demonstration plant converts coal to clean, gaseous fuel. Coal drops through the 135-foot hydrogasification reactor (tall foreground tower) where it is subjected to intense heat. As the gas is formed, it rises through the falling coal and is routed out the top to a purification stage.

The cost of such treatment, in commercial operation, is likely to be large. Another point is that demonstration of commercial practicability in power plants is a far more complex and costly operation than in cars. Obviously this hampers the work; and indeed we have seen that automotive emissions are apparently coming under control faster than those of power plants and industry.

Air Pollution Control

As a matter of fact, according to a report by the Council on Economic Priorities,* the utilities are lagging in installation of equipment that would reduce their emissions. In a study of 124 fossil fuel power plants operated by 15 utilities, it was found that only 35 percent adequately control particulates; 46 percent continue to burn high-sulfur coal or oil; and 81 percent have no controls at all for emissions of nitrogen oxides (a more subtle and recently observed problem; compared with a large effort now underway to control sulfur oxides, little research on nitrogen oxides is being carried out). As regulations become more stringent, and/or are enforced more strictly, the price of clean fuels will go up even higher, and more shortages will occur, or more utilities will turn to nuclear plants. Already, say the power people, too-strict regulations are causing a switch to other fuels, which are in even shorter supply.

* An independent, foundation-supported organization.

7

Other Pollution Problems

IF GASIFICATION OF COAL, or some other process of cleaning that fuel, is found to be economically practicable, then another problem, now merely simmering, will come to a boil. For then the mining of coal will achieve epic proportions, and strip mining, already producing more than half of our coal, will increase enormously.

Strip mining can best be understood by contrasting it with deep mining. In the latter, miners dig a tunnel down to where the coal is, then dig or blast the coal out, after which it is carried through the hole or shaft to the surface.

In strip mining, the "overburden" or covering earth is simply stripped away, the coal is dug up, and the overburden is (sometimes) put back in place. The main problem is that if great care is not taken in both the removal and replacement of the overburden, a number of environmental problems arise.

Good topsoil is constantly being made by nature, but it is a slow process, and in some areas it is only a few inches thick, particularly in drier western areas. Disturb this and the land may no longer be able to support vegetation; it becomes a true wasteland. Erosion, silting (washing away of loose soil), and acid drainage (a common problem in mining anyway) become severe problems. Those living "downstream," that is, in lower areas, may find the land and water around them silted, poisoned, and polluted to an extreme degree. Great gashes may be ripped out of beautiful, or at least natural, countryside.

It is said that some 2 million acres of American land have been defaced in this way. There are some regulatory laws, mostly weak ones; in some cases a small cash sum must be set aside by the operator for certain improvements, which the less scrupulous strip miners might take to mean replanting a few trees here and there.

Another problem is that as the more easily accessible veins are taken, more and more overburden must be removed to get out the same amount of coal. In 1946, for instance, 9 cubic feet of overburden had to be removed per ton of coal; by 1965, this figure had risen to 13. Thus we come up against the paradoxical situation of having to produce more coal—at a time when each ton produces more environmental disruption and when stricter environmental laws are being put on the books.

There is a growing sentiment for complete banning of strip mining. Yet consider: The economic advantages are obvious; *Time* magazine reports that in 1970 the United States burned 530 million tons of coal, of which 254 million tons were strip mined. To have deep mined it all would have required an additional $500 million. Some areas have been returned to a condition as good and productive as they once were, and in some cases improvements have been made. Swimming, fishing, and other recreational areas have been created. Desirable plant and tree species have been planted.

There is also a safety factor involved; deep mining is, after all, the most dangerous occupation in the United States. Stronger safety and health laws, which have been needed, have cut down on productivity.

So before we allow the pendulum to swing too far, perhaps we should set up strict laws prohibiting strip mining in some areas, such as the mountainous areas of West Virginia where reclamation is extremely difficult, but not in flat or rolling areas of the West where it is easier. Reclamation requirements should be made more strict, however.

The Greenhouse Effect

It is fairly well accepted that a small change in global temperature (either way) might start a stronger warming or cooling trend that could be disastrous over the long run. Among the fearsome possibilities are melting of the ice caps and a rise of several hundred feet in sea level, which would inundate a large number of the principal cities of the world, many of which are built on the banks of some body of water. At the other pole lies the specter of another ice age.

Of all man's activities, the one that is having the greatest effect on climate is the burning of hydrocarbons. But we are faced with contradictory indications.

On the one side we find that the combustion of fossil fuels, no matter how, or how efficiently, done, always produces a lot of carbon dioxide (see table on p. 68). The carbon dioxide in the atmosphere (from natural as well as artificial processes) produces what is called a greenhouse effect. In a typical greenhouse, the glass permits more of the sun's rays to enter

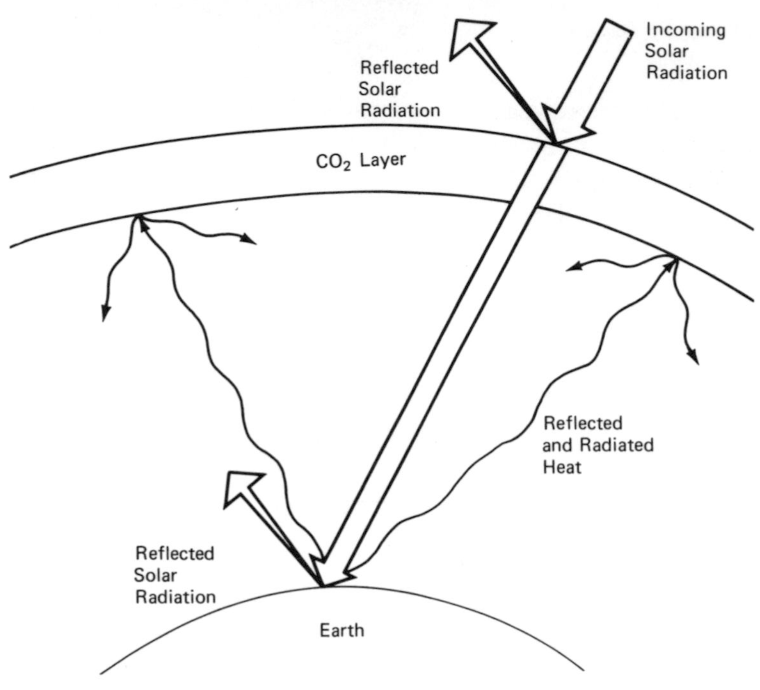

Greenhouse effect.

than to leave. In the atmosphere, the carbon dioxide layer does the same thing. The reason is a bit complicated, but has to do with the fact that the sun's rays consist of many types of energy—visible, ultraviolet, gamma, as well as radiant heat or infrared rays—to which the carbon dioxide layer is in general quite transparent. But most of the energy re-radiated back into space from the earth is radiant heat, to which the layer is less transparent. The atmosphere acts, in other words, as a sort of heat trap. As the amount of carbon dioxide thrown into the air increases, the greenhouse effect increases too.

The other, conflicting indication has to do with particulates in the atmosphere. These too are injected through both artificial and natural means. (In addition to combustion, other producers are volcanic eruptions, dust storms, and the like.) Such particles tend to act as a screen against incoming radiation, and thus make the earth cooler. It has been calculated that 50 million tons of particles added to the atmosphere and *retained* there (the particles do tend to settle out) could cause the average surface temperature of the earth to fall from its present 60° F. to about 40° F.—a temperature too low for most plant life to survive in. This is only about 10 or 20 times as much material as now is present in our atmosphere.

Professor Carroll L. Wilson of M.I.T., however, insists that we really know very little about the potential effects of an increase in particulates. In referring to a study of critical environmental problems carried out under the aegis of M.I.T. he says, "It has been widely assumed that particles had a net cooling effect and [that] if one kept in nice balance the net warming effect from CO_2 and the net cooling effect from particles, there was nothing to worry about. We discovered that the ignorance of the optical properties of fine particles was so great that they could either have a warming or cooling effect depending upon these characteristics."

Obviously we still have some homework to do.

But even assuming that particulates in the atmosphere do produce a cooling effect, we do not know which, if either, of the two effects (warming or cooling) will predominate. It may be that they *will* cancel each other out, in which case there will be no problem. Or it may be that they will set other, unbalancing forces in motion. We may even find that, if the warming trend predominates, we will have to deliberately pump particulates into the air. Wouldn't that be ironic!

In any case, we may find that man will simply have to stop the burning of fossil fuels, not because he is running out, but because of their impact on climate.

Radioactivity

The most obvious problems in the nuclear field are potentially dangerous radiations and nuclear wastes.

Although the actual release of radioactivity is truly minimal—a small fraction of the amount of radiation that is present on earth at all times—there are fears of release of large doses in case of accident. Though nuclear plants have had an excellent safety record up to now, this does not preclude the possibility of something going wrong in the future. As more and more plants are built, more and more radioactive material must be transported.* Although great care is taken, the material being shipped in extremely tough steel containers, breakage and leakage—and sabotage—are possibilities.

Equally difficult to solve is the problem of what to do with the radioactive nuclear wastes. A 1 million-kw fast breeder will produce about 10 cubic feet a year of fission end products. Although the volume of these wastes is far below that of equivalent coal-burning plants, the growth of nuclear power will cause a large total amount to be produced. Thus far these wastes are held in special, carefully guarded enclosures for a while to allow those parts of it that "cool" fastest to lose some of their radioactivity. Then a long-term storage must be found. It had been hoped that burying them in abandoned salt mines would be a final solution, but fears have once again arisen that water leaking in, earth movements, and other geological problems might cause venting or leakage of radioactivity into the air or water. The matter is still under

* Nuclear fuel lasts a long time but must eventually be reprocessed. This is normally done at special reprocessing plants, and the fuel must be shipped there somehow, as well as returned afterwards.

consideration, with storage in large surface vaults a current possibility. One suggestion has been to ship them off to the sun when a space transportation system becomes operational. But it would have to be an awful lot cheaper than it is now, or promises to be in the near future, before that becomes a realistic possibility. Maybe just orbiting them around the earth would be enough.

Thermal Pollution

All electrical generating plants and many other industrial plants as well produce great amounts of heat. As we have seen, out of 100 Btus of fuel burned, about 35 are useful, and the rest is waste. But, like garbage, it must be put somewhere; something must be done with it. It cannot be ignored. In general it is "shoveled" off into some body of water.

A 1 million-kw fossil fuel plant will heat up 30 million gallons of water per hour by 15° F. A nuclear plant operating at 32 percent efficiency will need 50 million gallons per hour, or more than half again as much. The difference lies mainly in the lower efficiency but also in the fact that the fossil fuel plant releases some heat up the stacks.

Already some 10 to 20 percent of the total U.S. fresh-water flow is used for power-plant cooling. This will increase sharply (estimates range up to 50 percent), not only because of the continued increase in power requirements but also because, as we saw above, nuclear plants are less efficient in their use of this water. An allied problem (faced by all water-cooled plants) is that this warmed water goes into our rivers, lakes, bays, estuaries, coastal waters, etc., all of which are the main incubating and growth regions for water life. Rising water temperatures can have devastating effects on this water life. Among the effects pointed to by environmentalists are fish kills, stimulation of less- or undesirable species of fish, an increase in the occurrence of disease in fish populations, and a decrease in the ability of the water to handle sewage.

It may, on the other hand, also be possible to use some of this waste heat—e.g., for "fish farming," warming soils in cold areas, and so on. In New York City and certain other areas some of the excess steam is piped to users for hot water and heat. In general this can only be done in tightly knit areas like city centers, which is exactly where power generating plants are no longer being built.

There are some other alternatives to shoveling the heat off into natural bodies of water, but they are more costly or more complex or both. For example, cooling lakes and ponds are possible. These are simply bodies of water into which the warmed water flows and is allowed to cool some-

Twin hyperbolic "wet" cooling towers, each 43 stories tall, near Sacramento, California, built to dissipate waste heat by recycling cooling water.

what before being put back into the main body of water receiving it (not always the one from which it was taken). But these eat up large amounts of land, which may not always be available. An industry estimate puts the land need at a minimum of about 1 acre per 1000 kw, which adds up to 1000 acres (1½ square miles) for a 1 million-kw power plant.

An alternative that may be seen more often in the future is the use of cooling towers. These are hollow, poured concrete, hyperbolic cylinders that may be 400 feet across at the base and 500 feet high. There are two basic types: one is a "wet" tower that absorbs a large part of the heat by evaporating some of the water; the other, a "dry" type, operates on the principle of the automobile radiator, with large volumes of air streaming past heated fins and carrying off the heat.

The wet type throws large quantities of vapor into the air and sometimes causes problems with fog, icing, even increased rain and snow in neighboring areas; the dry tower is more costly than the wet one and cuts down on the efficiency of the plant. So few dry towers have been built that it is not yet really known what effect all that additional heat will have on the atmosphere.

Nevertheless, if power plants must be built away from sources of water, this may be the price that has to be paid. Another problem with both types of tower is their enormous size. Environmentalists who are upset when a relatively small power line is put through an area (an example of "visual pollution") are not likely to be very happy to see several 500-foot towers climbing into the sky.

Siting

Once upon a time power companies could study various sites in secret, choose a site (there were plenty available and most communities were glad for the extra tax revenue), buy the land, get the necessary license, and build the plant.

Those days are pretty well gone, at least in the United States and most other developed countries. Today most communities say "Not here!"; even if the community leaders say okay, there will be individuals who take adanvtage of legal knowledge and maneuvers to block or at least delay construction of the plant. Ever more stringent environmental regulations are making whatever sites are available harder to put to use; and the number of approvals that must be obtained continues to climb, particularly in the case of nuclear power. Consolidated Edison of New York City must deal with five federal agencies, four state agencies, and 11 city agencies in the construction of generating plants. And in the case of one nuclear plant, it is said that 67 licenses were required, with delays and arguments necessary at each step.

And, finally, the number of good sites—meaning large areas near the population load but not too near, with plenty of cooling water available—has dwindled almost to zero. As a matter of fact, some of the plants that have been built in or at the fringes of cities will probably have to be phased out because of increasing population nearby.

Where, then, can power plants be built? A number of possibilities present themselves.

1) Offshore. The water problem is a problem mainly because our supplies of freshwater are limited. This is not true of ocean water, and the enormous size of the oceans means that they provide "heat sinks" of virtually infinite size. To take advantage of this, power companies are beginning to plan for location along the coastline. If plants are placed a mile or more out in the ocean, cooling becomes much less of a problem. The first such installation, a nuclear plant, is being planned for placement just under three miles from the coast, not far from Atlantic City. The plant is to be con-

structed at a "factory facility" in Florida and towed to the New Jersey site. (Other utilities, faced with inland siting problems, are also considering this approach, which includes possible siting in large lakes.) In this coastal design, the plant actually floats and moves up and down with the tide. But it is surrounded on the seaward side by a titanic breakwater (see illustration) built of interlocking concrete forms that are sunk in place. The power is delivered to shore through a buried cable. Offshore plants offer the advantage of less visual and thermal impact, no problems with land shortages, and, in seismically active areas, less chance of damage by earthquake; and the probably cooler water available from the sea bottom (though the water is only some tens of feet deep at this first site) can aid in the operating efficiency of the plant. Disadvantages are higher cost, higher maintenance due to the damp, salty environment, and problems with

Proposed pair of platform-mounted nuclear reactors moored within a common breakwater. Land is at left.

movement of men and supplies during fog and storms. Man-made island sites are also being investigated.

2) Underground. Due to the "dirty" nature of fossil fuel power plants, it only makes sense to talk of underground placement for nuclear plants. This has already been done in France, Norway, Switzerland, and Sweden. And in northern California a unit was placed partially underground. Such placement is probably psychologically reassuring to the general public, and minimizes visual impact as well. The elimination of external accidents caused by hurricanes and tornadoes and aircraft is also a plus factor. Balanced against this are difficulties in construction and repair and fuel handling, which add to the total cost of the plant. There is also, apparently, a historically low human productivity in underground work.

3) Energy centers. Undergrounding of power plants may make it possible to locate them close to or even within cities. Another approach suggests itself, however, the idea of nuclear "parks," in which one or more power plants could serve as the heart of an integrated energy complex, and in which all energy needs for a community are provided for on a cooperative rather than competitive basis. These might supply low-cost electricity, steam for heat and industrial processing, and the isotopes and ionizing radiation that are finding increasing use in medicine, agriculture, and industry. It is possible to think in terms of such an energy center serving as the nucleus for new and carefully planned cities in which much of the excess heat that is presently wasted can be put to use.* Clustering of nuclear plants might also make reprocessing of the fuel at the energy center economical and feasible, thus eliminating some of the need to transport this potentially dangerous material.

* See *The City in the World of the Future,* another book in this series, for further development of this idea.

4) Mine-mouth plants. In a similar manner, the economics of large-scale coal gasification points to the building of mine-mouth plants; these are power and/or gasification plants built near the source of the coal rather than near the population load area that is to use the energy. Shipment of gas is easier and cheaper than that of coal; and while shipment of large quantities of electricity over long distances is presently a problem, there are new methods on the horizon that may help.

The Heat Limit

It is important to understand that no matter what type of cooling is used, the waste heat and eventually all the heat generated must finally be dissipated in our environment. The importance of efficiency is that when less energy is wasted, much less need be generated.

Compared to our total heat budget (solar energy in versus heat radiated out into space), man's activities contribute a small addition. Currently the solar input is said to be some 20,000 times man's contribution,* which would seem to indicate that there is really nothing to worry about here.

There are two factors that must be considered, however. One is that while worldwide effects may not yet be a factor, local effects already are. Surface temperatures of urban areas are known to be several degrees higher than those of nearby rural areas. This "heat island" effect arises from blockage of air movement by tall buildings, by absorption and storage of solar energy in streets and buildings, and of course by man's industrial and other activities. In the Los Angeles area, for instance, the total amount of heat generated is about 3 percent of the incoming solar radiation, while in some highly

* The calculation is given at the beginning of Chapter 9, p. 105.

industrialized areas of the northeastern United States, northwest Germany, and southern Belgium, the figure is already up to between 5 and 10 percent.

As urban islands grow larger and perhaps merge, regional effects may begin to become an important factor.

After all, the solar input is, ultimately, the determining factor in our weather. Michael McCloskey puts it graphically: "At projected growth rates, by the year 2000 energy produced may approximate 30 percent of the solar input. Our population centers will turn into giant heat radiators affecting local climates. . . ."

The point is that we are becoming more and more capable—through sheer numbers, if nothing else—of unbalancing nature's systems. It may therefore become necessary to limit the population density and/or size of given large urban areas because of their potential negative effects on climate and earth's ecosystems. More research is needed in this area.

With better management of man's heat input, much of it could be routed to the oceans, which are large, cold, and generally available. It will be necessary to see whether large-scale dumping of heat into the oceans will have negative effects. Perhaps the water can be drawn via long intakes from great depths—the regions that remain cold because they are unmixed by surface waves and so are not heated by the sun. If the system is designed properly, the water can be heated only to the temperature of the adjoining surface waters. The discharge would thus create no troublesome temperature differences. There is even a possibility that this would bring forth nutrients from the ocean deeps. It would be a kind of artificial upwelling which, in its natural form, helps ocean life along certain coasts where it takes place.

Eventually, in the far future, man's activities could conceivably increase to the point where total heat would be a problem. The final limit to man's activities may take place when the seas begin to boil! That may take a while yet.

Fortune magazine reports that "scientists and technicians, in a frustrating search for some way to describe the changes that their work on nuclear energy portends, speak glowingly of air conditioning Africa and heating the subartic." *

This was written in 1967. Most of us now shudder at the very thought, for both processes would release enormous quantities of heat into the atmosphere.

But we are speaking about the future. And it is conceivable, though only barely so, that the heat removed from tropical regions could somehow be piped to colder regions! Obviously, even if this would not add to the total heat load of our earthly activities, the redistribution would still have an enormous effect on the ecosystem of our earth.

A slightly more realistic idea along this line is that of towing great icebergs from the polar regions to the warmer climates. Not only would this be a kind of water conditioning (along the idea of air conditioning), but it might supply freshwater to some drier areas as well.

We see that a major need is some form of energy production that does not add to our planet's heat load.

Up to now we have concentrated on the problems caused by energy production and use, and we have mentioned some of the needs. Let us now look at some solutions.

* March 1967, p. 121.

IV. TOMORROW

8

Geothermal Energy

A VOLCANIC ERUPTION can assault the senses: Enormous quantities of hot gases, steam, ash, and dust shooting into the air; molten, red-hot lava curling down the sides; a sulfuric smell; earth-shaking thunderous roars. All of it spells spectacular. All of it testifies to the enormous forces locked up within the earth. A volcano is merely one example, though one of the more spectacular ones, to be sure, of the heat and pressure that are contained within the earth.

I found an interesting confirmation of this a few years ago when I took a walk to the top of Mt. Vesuvius; this is the volcano whose eruption in A.D. 79 buried the thriving resort towns of Pompeii and Herculaneum. Here and there on the sides of the volcano were a number of shallow holes which were hot enough to ignite a piece of newspaper dropped into them!

Even aside from its shock value, this is interesting on several counts. It reminds us that heat is associated with volcanic activity; that it is not very far down below the surface of such an area; and that rock is a very good insulator of heat (since the region seemed no warmer than neighboring areas).

Volcanoes and volcanic areas are nothing more than places where the hot rock deep within the earth has, for one reason or another, risen to, or close to, the surface. If this has taken place in any recent era, geologically speaking, then the region just below the surface has probably remained warm and may contain the necessary ingredients for the natural production of steam and/or hot water.

In many areas of the world, naturally occurring hot-water springs have been thought to have curative powers and have been used as medicinal as well as recreational baths. Hot springs were used by the Romans, for example, in what is now Hungary, 2000 years ago. It is interesting to note that right after World War II an extensive geological survey was made in Hungary to search for a long list of mineral resources, very few of which were found. Great emphasis was placed on developing some small deposits of oil and coal (Hungary is not well blessed with fossil fuels), but the hot springs were quite ignored as a possible energy source. Even today, except for some small use as a source of heat in new housing developments and in a few factories, this natural heat has not yet really been put to use. As in many other countries of the world, however, interest is rising rapidly in this generally ignored resource.

Both steam and hot water can be put to use, though in different ways, to produce electricity. If conditions are right (see accompanying diagram), it may be possible to drill a hole and find a "steam well." While this may sound less exciting than finding an oil well, the future potential may be as great. And drillers are beginning to realize this more and more. Indeed, oil companies are among those now searching for geothermal sources.

The conditions required include the following: A source

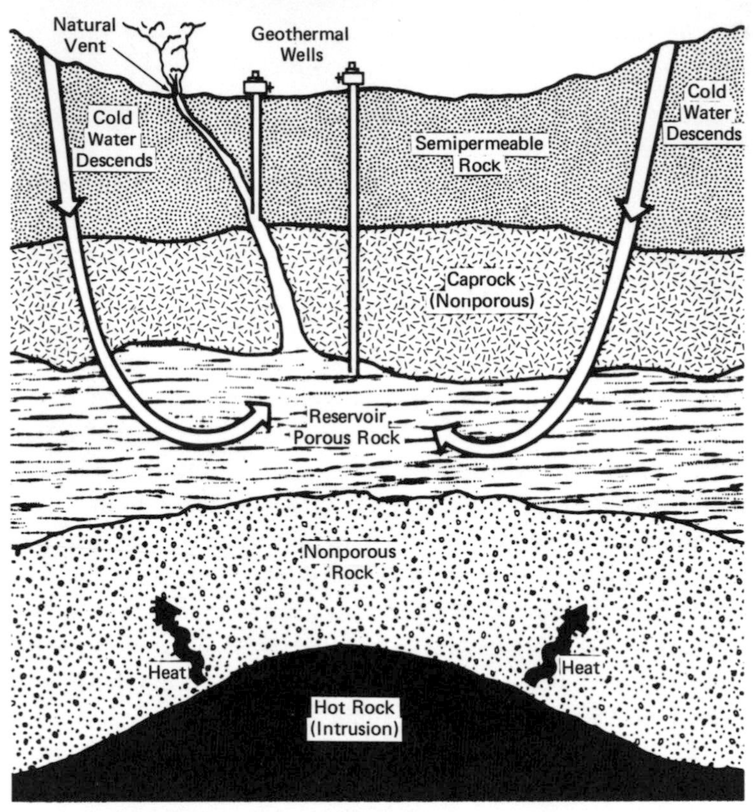

Geothermal steam surfacing from hot rock intrusion.

of hot rock, brought up from the earth's depths by some geological activity in the recent past. This is the heat source, believed to be maintained by radioactivity. Above this there must be a reservoir of water, which is free to circulate, contained in some form of permeable or fractured rock. The next higher layer would be a kind of less permeable caprock to hold down the water and hold in the heat. There may also be a layer of nonporous rock between the intrusion and the reservoir which conducts the heat upward.

Being deep down, the water is under great pressure and so the temperature can rise above its normal boiling point; water temperature may be up in the range of 500° F. If a natural opening or vent exists, the water may move into it. With pressure released, it will flash into steam and erupt. Or a well can be dug to tap into the porous layer, and the resulting "dry steam" (no water mixed in) can be used to power a turbine.

The first such application took place in Larderello, Italy, beginning in 1904; the steam field has been in continuous operation ever since, and has now been built up to a capacity of 380,000 kw.

Other existing geothermal power plants are found in Mexico, Iceland, New Zealand, Japan, the U.S.S.R., and the U.S.

In the United States, the only producing area is in a region called The Geysers, about 90 miles north of San Francisco. Put into operation in 1960, it now has a generating capacity of 192,000 kw, with more planned. It is also the only geothermal power plant in the world that is being run by a private company, which is interesting. Research and development costs can be a strong detriment to development of a new energy resource. The fact that private enterprise thought it worthwhile shows that a high economic incentive exists. Now that pollution factors and fossil fuel shortages are becoming more important, it is likely that more such development will take place. (It must be kept in mind, however, that air pollution *is* present in geothermal energy. Hydrogen sulfide is the main problem.)

Naturally occurring dry steam is relatively rare, and this may be one of the reasons that geothermal power has traditionally been thought of as a minor resource. (Others are that both pollution and shortage of fossil fuels were not considered problems until recently.) It has been found, however, that another type of geothermal source is far more common. So called wet steam fields may be 20 times more common than the dry type.

In dry steam fields geological conditions combine to produce a situation where temperatures are high and pressures low, permitting boiling down in the reservoir. The only treatment necessary is grit and rock removal. The low-pressure natural steam can then be fed directly to low-pressure turbines.* In wet steam fields reservoir pressures are higher and boiling cannot occur underground. When a well is drilled, however, the hot water flows into it and flashes into steam on the way up. What emerges is a mixture of steam and hot water.

The magnitude of these pressures was dramatically demonstrated in 1966 at Cierro Prieto, where the Mexican government was constructing a large geothermal steam plant. In September of that year workers were drilling a well when a 5-ton valve was tossed off the well as if it were a cork. The steam column spurted almost 500 feet up, and the roar it produced was said to be deafening a quarter of a mile away. Seven workers were injured.

Among the problems faced in the geothermal field is the fact that, particularly in wet fields, a steam-condensing plant is needed, meaning that cooling water and/or cooling towers are necessary. Also, the steam and water usually contain a high proportion of minerals. On the other hand Professor Robert W. Rex, who directs the University of California Geothermal Resources Program, would turn a problem into an advantage. He believes that because the minerals are in solution at this point, this may provide an important way of obtaining minerals in the future. The point is that if a gas

* Thus large volumes of steam must be handled, which presently limits the practical size of individual generating units to about 55,000 kilowatts.

(such as steam) at a low enough temperature is compressed, it will liquefy. This is what happens in steam fields. But for every gas there is also what is called a critical temperature above which it cannot be liquefied, no matter how much pressure is exerted. Under certain circumstances, the temperature may rise above the critical temperature, producing a supercritical state. Dr. Rex believes that "The chemistry of the supercritical state in geothermal systems is one of the exciting new frontiers of modern geology, and we are finally developing the technology to study it in nature. This near-supercritical and supercritical environment is especially exciting as the environment of formation of metal ore deposits. One of the new technological frontiers of the next 30 years should be the study and development of ore recovery from *live* geothermal ore solution."

An important economic aspect of geothermal energy, and particularly that of wet fields, is the possibility of putting the hot water to use as well. At the moment, however, this highly mineralized effluent is a problem, for it tends to rapidly deposit in the pipes; after a while the whole system may have to be redrilled. At the moment, too, the effluent is something that must be disposed of. In the Wairakei field in New Zealand, the effluent is run off into a river, an arrangement that would probably not be permitted here. Another approach which is being looked into is to reinject the effluent back into the reservoir, which will help prevent a potential future problem, the subsidence of the earth in the area as a result of removing the water from the interior.

But the Wairakei field, begun in the 1950's, was still an early experiment. Present approaches look to multiple use of the water and steam. Another possibility, particularly in water-short areas, is to use the heat from the water to distill out the "pollutants," thus creating additional supplies of freshwater. At a wet steam field recently found in Chile, it is

BELMONT COLLEGE LIBRARY

View of production field at Wairakei, New Zealand. Geothermal steam is being discharged from waste hot water silencers.

planned to combine three uses of the resource. The steam will be used to produce electricity. Hot water (which can also be obtained by condensing the steam) will be put through a desalination plant to produce freshwater. And the effluent from this process, in which the mineral content is even more concentrated after freshwater is extracted from it, will be fed to evaporation ponds, from which minerals will be extracted. If desalination proves feasible, large-scale development may proceed in the Salton Sea and Imperial Valley areas of California, where wet fields have recently been discovered.

And, as we mentioned earlier, the hot water has been used for heating and industrial processes for a long time. There is no reason why this use too cannot be built into wet steam applications. In Hungary, for example, the use of hot water for home heating is said to cost only about one-quarter of equivalent fossil fuel systems. In the United States, the same thing is being done on a small scale in Klamath Falls, Oregon, and Boise, Idaho. The U.S.S.R. is said to be producing a system which uses the heat energy of hot water for both heating and cooling.

Thus far a total geothermal generating capacity of about 1 million kw has been developed worldwide. At the present rate of development, says Joseph Barnea, director of the United Nations Resources and Transport Division, production of electric power from geothermal sources will be quadrupled by the end of the 1970's. In the U.S. this will be aided by the 1970 Geothermal Leasing Act, which will make available millions of acres of promising land for geothermal prospecting.

The total potential, as with all energy sources, is extremely difficult to figure. But estimates have gone as high as 10 billion kilowatts for the western United States alone (which happens to have a rich supply of natural steam). On the other hand, David E. White of the U.S. Geological Survey figures that because these resources are depletable, if averaged over 50 years the annual production would come to only 60 million kilowatts. Clearly we still have much to learn in this area.

It is also worth noting that a geothermal well system can bring only about 10 percent of the heat from a reservoir up to the surface. Combining this with the 10 percent efficiency of geothermal plants, we find that over-all conversion efficiency is only about 1 percent.

Dry Heat

On the other hand, geothermal potential will become almost infinite if a new idea put forth by scientists at the Los

Alamos National Laboratory in New Mexico proves practicable. It is known that the temperature of the earth rises everywhere at a rate of some 10° to 20° F. per 1000 feet, and sometimes more. In large parts of our western regions, temperatures of 500° to 600° F. are believed to occur within the top 20,000 feet. The new idea is simplicity itself. Drill two holes down far enough to reach hot rock. Into one hole you pour cold water; this circulates through porous or fractured hot rock, and comes up through the other hole as steam!

If the rock down below is naturally porous, a natural reservoir might occur. But by a technique called hydraulic fracturing, a zone of fractured rock can be created around the pipe. Water is forced down the pipe, which is perforated at the sides, at pressures as high as 7000 pounds per square inch, fracturing the rocks and creating the necessary porous zone. Between 2 and 6 miles down, says Dr. Morton C. Smith of the laboratory, there are enough dry geothermal reserves of heat "to satisfy this country's electrical energy requirements for several thousands of years."

The nicest part is that we are not restricted, as we are in the other two types, to areas which have natural reservoirs of dry or wet steam. There is no region that is not underlain by hot rock.

The water won't even have to be pumped. The water going down will be cold and dense; that coming up will be hot and therefore less dense. It will as a result be "pushed" up by the colder, denser water going down.

There are a few complications, as you might expect. Present heat requirements indicate that a 600° F. zone would be required. For a nonvolcanic area, this would require drilling as deep as 12 miles, which is far beyond any practicable present technique. Further work may bring this requirement down to 340° F. And in any kind of volcanic area, of course, even the higher heat can be reached much more quickly. A demonstration project is being planned (assuming funding can be obtained) for an area near the rim of Jemez Caldera, an extinct volcano in New Mexico near the Los Alamos Laboratory, that would require drilling down only 15,000 feet. This is well within the capabilities (about 25,000 to 30,000 feet) of conventional oil-well drilling equipment. Beyond that the drill stem begins to wind up like a twist drill bit.

At Los Alamos, a new drill is being developed that may overcome this problem and permit even deeper wells to be drilled. The heart of the new device is a heating element that is hot enough (3000° F.) to melt rock (melting point about 2200° F.), automatically leaving a half-inch thick glass lining.

Another approach would be to explode a small nuclear device 1 or 2 miles down and inject water into the cavity. The resulting hot water and steam could be used in a closed, circulating system (to prevent contamination) and then returned to the cavity. There are, however, dangers of leakage from the cavity itself, and even venting from an accidental or natural blowout. It is thus unlikely that we will see development in this area, at least for a while, even though a bill was introduced in the U.S. Senate in March of 1971 to authorize the A.E.C. to study the idea.

(It may also prove possible to find a large-scale use of nuclear wastes by putting them to work as the heat source. Underground "chimneys" would be created, into which the wastes are deposited. Water would be pumped through to control the temperature, and would also be drawn off, or sent through a heat exchanger to produce usable steam.)

At the moment the three different types of geothermal power we have discussed—dry steam, wet steam, and hot rock—are successively more expensive. On the other hand, the power obtainable is successively more plentiful. There is some comfort in knowing it is there, even if it is still hard to get at.

Another, even more distant possibility, is to somehow make use of the thermal gradients (heat differences at different levels) of the oceans; a scheme to accomplish this, suggested by Professor Clarence Zener of Carnegie-Mellon University, is illustrated.

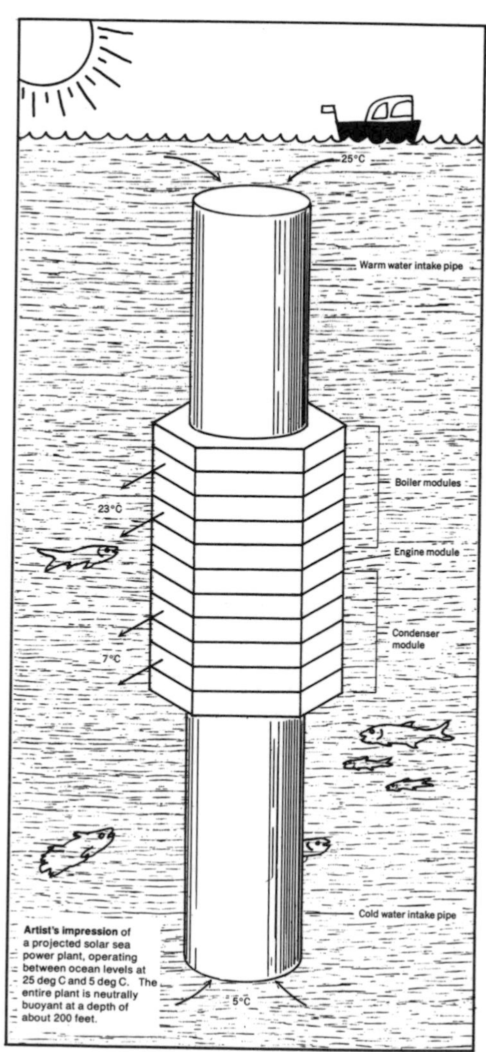

Projected solar sea power plant.

Artist's impression of a projected solar sea power plant, operating between ocean levels at 25 deg C and 5 deg C. The entire plant is neutrally buoyant at a depth of about 200 feet.

9

Solar Energy

THE EARTH RECEIVES from the sun each year 1.6×10^{18} kilowatt-hours of energy. Total human energy use (excluding muscle power) is some 7×10^{13} kw-hr per year. Dividing the first figure by the second tells us that the earth receives more than 20,000 times as much solar energy as man presently consumes from all other sources.

Of course much of this energy goes into wind and wave, as well as the evaporation of ocean and freshwater that eventually produces mountain streams, rivers, and waterfalls; * and, as we mentioned earlier, it provides the energy for photosynthesis, without which animal life would be impossible.

The sun, in other words, is the prime mover in the production of practically everything on earth—our weather, our food, our fossil fuels, our climates, and so on. But even so,

* The sun does an enormous amount of work. Each year, for instance, it evaporates and raises perhaps 100,000 cubic *miles* of moisture from the earth's oceans, rivers, lakes, and streams.

there is plenty left unused and simply radiated back into space.

Narrowing our scope to the United States, we find that even its enormous energy needs for a full year could be met from the sunlight that falls on it for a single day. Or, turning this around, all our energy needs could be met from the energy falling on perhaps 1 percent of our land surface. This energy is free, nonpolluting, and we don't have to worry about running out of it.

Putting the sun's energy to work is obviously not a new idea. We have tanned our skins, dried our fruits, and warmed our bones in its rays. But these uses are like drops in the ocean compared to its potential use for electric power, as well as for hot-water and space heating.

Why then do we seem to be having so much trouble getting usable power from it? Ignoring economic considerations for the moment, there are two main drawbacks to solar energy. First, while the energy that falls on the earth as a whole is continuous, that which reaches a given spot on earth is not, earth's rotation and intermittent cloud cover being the main interrupters.

Secondly, what does arrive is of low intensity. Allied with this is the fact that present methods of storage and transportation of heat energy (as well as electrical energy) are extremely inefficient.

Collecting Solar Energy

There are two basic ways in which the sun's energy has been, or can be, collected. In the first, exemplified by photosynthesis and sun tanning, the sun's rays simply fall on a surface and in some way are put to work. On a clear day the solar energy falling on a square meter of surface amounts to something over 1 kw. Spread out over an acre it amounts to more than 4000 kw.

The majority of the sun's energy comes to us in the form of light, not heat. Depending on the characteristics of the surface on which the light is falling, a certain amount will be absorbed and the balance reflected back into space. For light, the relative darkness of the surface will be the determining factor. The darker the surface, the more energy will be absorbed.

But then a strange thing happens. If the light energy is not used, as it is in photosynthesis, then the energy that is absorbed is converted to heat. Some will leak away, but if absorption takes place at a faster rate than the heat is lost, the surface gets hotter and hotter. At a given temperature, the surface begins to radiate energy back into space; but the energy now goes off as heat, or infrared radiation.

Now, we saw that the efficiency of a heat machine depends on the working temperature of the device. So in order to make the surface as hot as possible, we must make it a good absorber in the light spectrum and a poor emitter in the infrared. A simple black, dull surface does a fair job of this. Some newly developed surface coatings are helping.

Concentration

The second general approach to collection of solar energy is exemplified by the schoolboy who concentrates the rays of the sun with a magnifying glass and by so doing chars paper or actually sets it on fire. Clearly the rays of the sun are not hot enough to do this by themselves. What the magnifying glass does is to collect the rays from a broader cross section and concentrate them into a small area. A concave mirror can be used to accomplish the same result. (In a

Solar cooker (supplied by Edmund Scientific Corp.) cooking two hot dogs in author's back yard.

reflecting telescope a concave, parabolic mirror is also used, but the objective is to accumulate as much *light* as possible.) A great deal of research has gone into devising workable solar concentraters. Such a device can produce a temperature as high as 350° F., hot enough to boil a quart of water in 15 minutes. The photo shows a unit that the author and his daughter tried out in their backyard. It worked.

Obviously larger mirrors can produce higher temperatures. In the mountains of southern France, 3500 small mirrors are arranged in the form of a parabola. At the focal point, where all the collected rays are focused, the temperature goes up as high as 5400° F., thus making it a useful instrument for research in that it produces clean, high heat.

Electric Power Production

Can solar energy be converted to electricity? The answer is an unequivocal yes. Is it worth doing? This question is much harder to answer.

First of all, it has been done many times. It is being done even now, but only on a minuscule scale, and only for applications where it offers some strong advantage over other methods. One way is through photocells, similar to those used in photographic exposure meters.

Top: Small DC motor powered by solar cell spins cardboard circle. Bottom: When sun's light is blocked off, the motor stops.

Solar-powered seismograph station.

Photocells are being used to power a group of seismograph stations set up along the Gulf of California to monitor earthquake motions of the Gulf's floor. On the roof of each station, 540 cells charge batteries that can store power for up to 3 days. Thus the station needs sunlight at least 1 day out of every 4. In space probes, where the sun is always available and at full strength, the photocells receive continuous power, and the "fuel" is free.

While the photovoltaic cells, such as the silicon solar cell, work well, and can presently convert some 12 percent of incident solar energy into electricity, they are enormously expensive. Space power plants have run on the order of $200 to $300 per watt. At present the over-all cost of a silicon cell battery combination is about 1000 times higher than conventional electric power.

A gallium arsenide cell made by I.B.M. is said to operate at an efficiency of 18 percent; continued research may produce efficiencies as high as 22 percent, which would of course help bring the cost down. Better production techniques and greater quantities would too. It is possible that cost reductions of 10, 100, or even 1000 times per watt are attainable. Perhaps we will yet see a farm or garden tractor being run from solar cells in the umbrella that shades the operator from the sun.

At the moment, however, power cells are surely not a practicable idea for power generation on a large scale for any application. Another approach may be, however, one which puts to use the standard water/steam cycle, but replaces fossil or nuclear fuel with solar heat. A system proposed by Aden B. Meinel, director of the Optical Sciences Center at the University of Arizona, and his wife, Marjorie, may be able to do this. They write: "Thermal conversion of solar energy has not been particularly successful, and it is clear that the resultant low efficiencies, less than 2 percent, are due to low operating temperatures . . . so one avenue for increasing efficiencies is to increase the input temperature of the thermal conversion cycle."

With an eye toward interfacing with current generating-plant conditions, they chose to try for 1000° F. at 1200 pounds per square inch of pressure. To accomplish this, they proposed that the sun's rays be concentrated first by glass or plastic lenses, then again by a reflecting collector which directs the rays onto a pipe. Running through the pipe

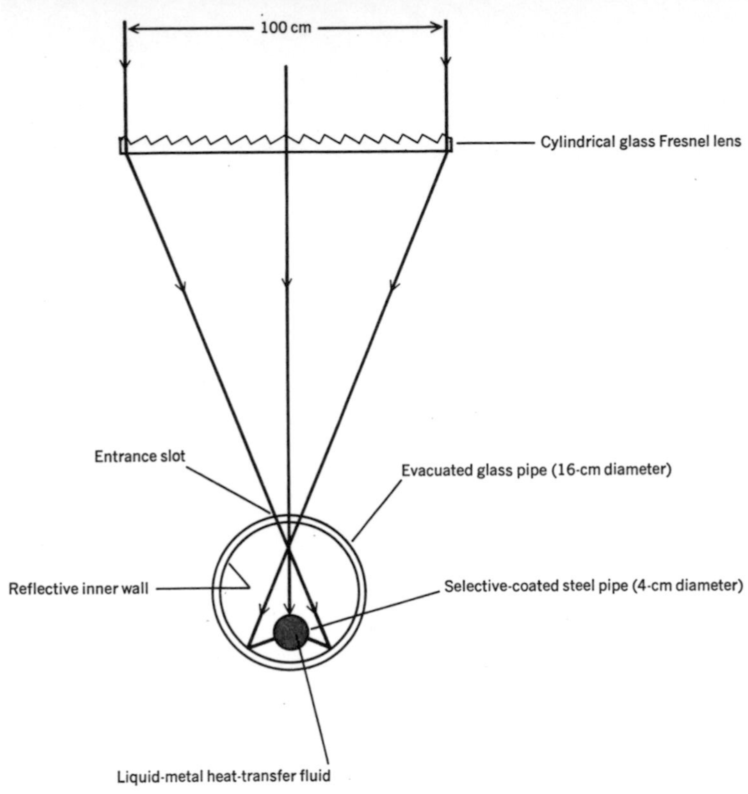

Energy-collecting element, shown in cross-section. 6 to 8 of these mounted on a supporting structure could become a "solar module."

would be a sodium-potassium mixture (as in the LMFBR), which would transfer the heat to an insulated chamber that could store it for at least a day. As shown in the diagram, heat extracted from the chamber would run a conventional steam-electric power plant. Hoped for over-all efficiency would be about 30 percent.

A very different system has been proposed by A. F. Hildebrandt and G. M. Haas of the University of Houston. Here the radiation striking a square mile of surface would be reflected and concentrated onto a solar furnace and boiler at the top of a 1500-foot tower. Boiler temperature would reach some

2500° F. and would be converted to electricity by magnetohydrodynamics. It is estimated that a total efficiency of 20 percent can be realized.

With such an efficiency, a 1 million-kw plant would require some 13½ square miles of collecting area. To satisfy the total electricity generating requirement of the U.S., which now stands at about 350 million kw, would therefore require 4728 square miles, or just under the area of Connecticut.

This sounds like it puts the whole thing out of the realm of reality. On the other hand, as the Meinels point out, we

Schematic diagram of a thermal conversion system.

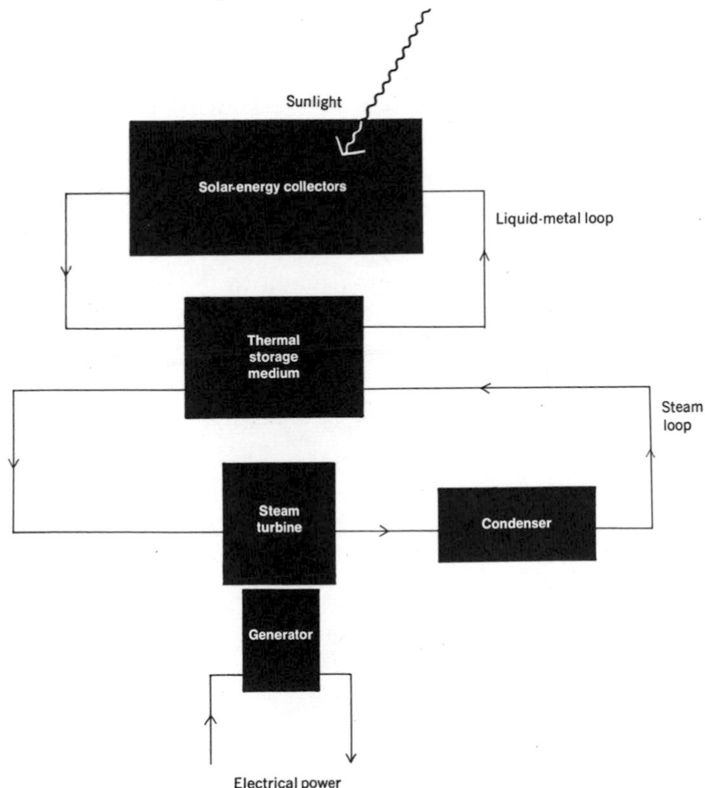

don't hesitate to devote some 500,000 square miles, or almost 14 percent of our land, to the growing of food. While it is true that we as living organisms are wholly dependent on food for our existence, the same might he said of the importance of energy in our modern societies. Further, our food supplies only about 1 percent of our total energy needs.

A Space System

If we should decide that solar energy is the way to go, and at the same time that we cannot afford to devote large tracts of land to its collection, the ever-working mind of man has already provided an alternative. Peter E. Glaser of A. D. Little, Inc., an engineering firm, would put the collector up in space. A satellite located at a height of 22,300 miles is essentially stationary in space with respect to the surface of the earth. That is, it remains above a given point. A panel of solar cells would therefore have the advantage of being able to collect energy almost 24 hours a day. It would be converted in space to microwave energy and beamed to a receiving antenna on earth.

Glaser calculates that a 10 million-kilowatt station, large enough to supply the electrical needs of New York City, would call for a solar collector 5 miles square, and a receiving antenna on earth 6 miles square (36 square miles). This compares favorably with, say, the Hildebrandt/Haas scheme (135 square miles of collecting area), and has the additional advantage of providing continuous power. Cost would run at least $500 per kilowatt, or more than twice the cost of a nuclear power plant, even assuming that the space shuttle now being developed by the National Aeronautics and Space Administration could be used to transport the materials up into space.

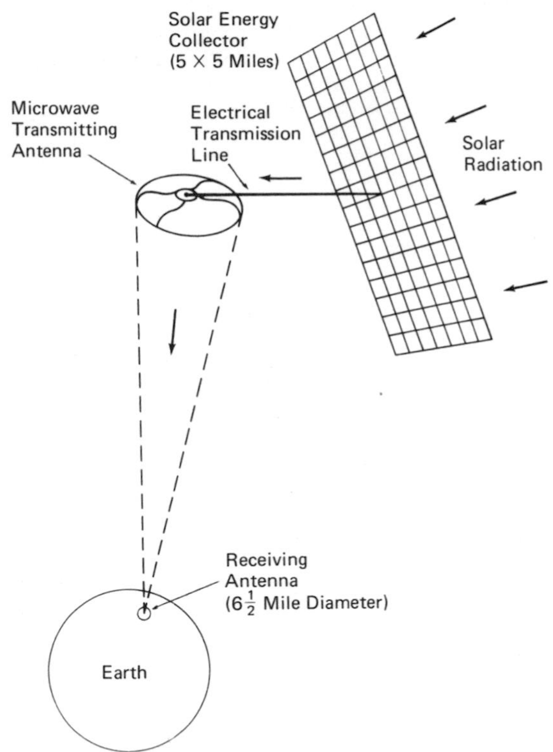

Proposed solar energy collector.

Additional advantages claimed by Glaser: No problems with deposition of dirt and dust on the collectors, or erosion by wind, water, and weather, nor with wind effects on the structure of the collector; the input is constant, instead of variable with the changing angle of the sun, cloud cover, and day/night cycles in an earth-based system; and, finally, 80 percent of the thermal waste is dissipated in space.

Space and Water Heating

Somewhat more straightforward and easier to achieve is the use of solar energy for direct heating. This heat can be used in many ways. It has long been used, though not on a large scale, for desalination of sea water and for evaporating sea water to obtain salts and chemicals. Although it has become more economical to mine solid salt, large quantities of salt are still obtained by evaporation of brine in vast "salt pans." Each pan produces about a 4-inch depth of salt crystals each year. Although a million tons of salt a year are still produced this way, this is clearly a very specialized use. There is greater potential in solar stills, which would produce freshwater (if a good system could be found) from salt water in hot, coastal desert areas.*

Much more important in an economic, and perhaps ecological, sense would be to put the sun's energy to work for space and water heating.

There are two major reasons for this. First of all, something like 20 percent of the nation's energy consumption goes into these uses, almost all of it coming from gas and oil which, as we have seen, are coming into a short-supply market. Twenty percent of 70,000 trillion BTUs is surely worth considering as a strong potential market.

But there is more to the story than this. Fully 10 percent of our energy use goes into residential space heating. A typical oil- or gas-heating furnace can be 70 to 80 percent efficient—when it is in good adjustment. Unfortunately, furnaces tend generally not to be in good adjustment, with the result that efficiencies may be as low as 40 to 50 percent, and in water heaters as low as 30 percent.

Solar furnaces, it has been shown, can be relatively efficient and can convert more than 70 percent of the incoming solar energy into usable heat. In general, temperatures no higher than 400° F. are sufficient, which means that the solar energy engineer needn't get involved with concentrating the radiation.

* Strangely, most of the world's deserts are near the sea.

Already several million rooftop solar heaters are providing hot water in places like Israel, Australia, and Japan. House heating is a bigger and, hence, more difficult job. In general what is done is the following. The house is built with a large, dark collecting surface tilted and oriented to face the winter sun. Air, or more likely, water, is circulated across the heated surface and the whole thing is covered with one or more layers of glass or plastic. These easily pass the incoming energy from the sun but tend to keep in the converted energy that tries to escape in the form of infrared rays. This is the greenhouse effect we discussed earlier.

The fluid, after collecting the sun's energy, is sent through some sort of well-insulated storage section, which can be a large tank of water, dry gravel, stones, or sealed cans of molten salt. A mixture of chemicals called Glauber's salt, for instance, remains solid until the temperature rises to 90° F. Above that it melts, and in so doing absorbs large amounts of heat. As long as the salt remains in the melted condition, the heat is held in what is called the latent heat of fusion, which can then be released as needed as the salt solidifies again. The basic idea is the same as the extra energy needed to freeze water while the temperature remains steady.

The advantage of the salt is, first, that the "freezing" takes place at a much higher, and thus more useful, temperature than with water; secondly, it has a higher latent heat of fusion, which means it can store more heat. Large amounts of the salt are needed, however, the actual amount depending on the size of the house, the average temperature outside, the size of the collecting surface, and the number of days of sunshine. A general figure might be 10 tons of rather expensive salt to store heat for a week.

In colder areas a supplementary heating system is also a necessity, and it may have to be almost as large as that of a nonsolar-heated house if one wants to be sure of not running out of heat.

A number of homes have nevertheless been built or adapted to utilize solar heat, though mainly in warmer areas, and normally without highly expensive approaches like the Glauber's salt approach. In Melbourne, Australia, a home is being both heated and cooled with the aid of 50 tons of crushed rock in the form of a wall 48 feet long, 6 feet high, and 4 feet wide. In the winter, the rocks are heated through a glass pane on the outside wall, and air is circulated through the house when needed. During the summer, the rocks are cooled by circulating night air through them, and hot daylight air is cooled by circulating it across the rocks during the day.

While this sounds like a rather unwieldly approach, it may be that if energy supplies diminish rapidly enough, we will find ourselves forced to use such extreme measures. The same holds for the pollution picture, for it turns out that all other forms of energy production have some form of pollution connected with them. Even solar energy may produce problems with heat balance, and the conversion process is not completely innocent either. Putting a large installation into a relatively delicately balanced desert environment could have destructive effects. But solar power is surely one of the least harmful over-all. Or it may be that antipollution laws, combined with generalized shortages, will drive the cost of all other types of fuels way up, in which case such measures as the Australian or even the Glauber's salt approach may not seem so extreme after all.

Should We Do It?

Professor David J. Rose of M.I.T. has calculated what it

might cost to go to an all-solar electric-energy technology: Our total electric bill might triple to some $75 billion. This extra amount, he suggests, is roughly equivalent to the cost of a mild and temporary recession. "Thus," he says, "we could quite well, if there were a societal decision to go for it, solve in a narrow sense a substantial number of problems connected with electric power production."

But as with so many other technological "fixes," this one carries dangers with it. While the cost of conversion could undoubtedly be borne, other undesirable changes might result from it. If energy becomes much more expensive, we would probably see a movement from present energy-intensive processes to more materials-extravagant ones, causing or intensifying materials shortages. The mining of low-grade mineral ores would have to be abandoned, leading to more rapid exhaustion of such materials as zinc. If we did this and other countries didn't, our products would become more expensive than theirs and our competitive position would be weakened. "I do not know," Dr. Rose adds, "and neither does anyone else, just where things would settle out. . . ."

In the near future the most favorable regions for development of solar power are in the desert areas, and particularly those within about 35° north and south latitudes. Improvements in storage and transmission will be needed, if the power is to be useful on a continuous basis, and if it is to provide power where it is presently needed. On the other hand, wide-scale application in desert areas might provide the basis for building new cities, based on new technology, and experimenting with new ideas.

10

Wind and Water

BY THE EIGHTEENTH century man had developed a wide variety of machines to help him—in growing food, manufacturing textiles, and so on. With the exception of a few steam engines used for pumping the water out of mines and for running mills, the motive power was, as it had been for centuries, wind, water, and animal power.

With the further development of steam, internal combustion, and electrical power, these began to decline in importance. But today, with fuel shortages impending, we must reconsider all options, and it is possible that we will be using more of the ever-present, pollution-free energy offered by wind and water.

It has been estimated that falling and/or rapidly flowing water could provide man with anywhere from one-third to three-quarters of his present energy needs, though at present he utilizes only about 1 or 2 percent of the total available. If all the world's tides could be put to work, they might provide half his needs. And winds might be able to supply all his needs. In this chapter we explore the promise and problems of putting these energy forms back to work.

Wind

Aside from sailing ships, the best known application of wind power is the windmill. In both cases the wind is being used directly as motive or driving power. Today, if put to use, the wind's energy is more likely to be used to power a generator.

In the Eastern regions of the world, where prevailing winds are common (steady at least in direction, if not in velocity), the windmill could be built as a solid unit. In the Western parts of the world, where winds are variable in both direction and speed, it had to be built so that the part containing the vanes, or even the entire mill, could turn so that it would always face the wind. Thus the windmill as it developed in the West was actually quite a complicated affair. It found widespread use in northwestern Europe, and particularly in low-lying Holland, which suffered from a lack of water power.

As with solar power, wind power has two serious drawbacks—low density and variability. In the latter sense it may be an even less reliable source than solar power. What does one do on calm days? In years gone by, users were content to wait it out. We are not likely to stand for that today. Standby machinery can be used, but it requires an additional investment. A wind-catching machine, whatever form it might take, could be combined with some form of storage device. Small home units might be used to drive a generator which could charge the batteries for an electric car or other use. Additional possibilities are suggested in Chapter 13.

It is also possible that transmission of energy will become so efficient and cheap that, with the help of modern sensors and computers, excess energy produced in one place during the night might be transmitted to another area thousands of miles away that is in a daylight, peak period. Such a capability would also make it possible to put wind units in those places which are unpleasant, and hence unpopulated, perhaps *because* of high or steady winds–such as mountaintops, for example. At the moment, however, there is not a great deal of interest in wind power, although experimental machines have been built in France, Denmark, and the United States (see photo).

Hydroelectric Power

When power needs were far smaller, the energy contained in the flowing water of a nearby stream was often enough to satisfy those needs, and so there were water wheels that were turned in this way. For more power, however, the energy of falling water is needed. In a modern hydroelectric power plant, the falling water is passed through turbines, turning them and thus generating electricity.

If there is flowing water, but no drop, this can be corrected by building a dam if the water is flowing through a valley. The dam holds back the water until the reservoir behind it fills up (the valley walls become the sides of the lake); then the water is allowed to "fall." Dams may also be used for flood control, irrigation, and water supply.

Mountainous countries like Switzerland and Norway use proportionately more hydroelectric power than we do. In the Pacific Northwest region of United States, however, hydroplants make up more than 90 percent of the total installed electric generating capacity.

Forty-foot-tall wind generator designed by Thomas E. Sweeney uses unusual blades.

Hydropower provides some definite advantages. It operates entirely without fuel; the original energy comes from the sun. Thus there are no combustion products, or any other wastes. The process is almost 100 percent efficient, so very little waste heat is generated. And the electricity so produced is usually cheaper than that of any other forms. This is sometimes offset, however, by the need to transmit the energy over long distances, for it often happens that the hydropower sites are not near the places where the power is needed. A project nearing completion at Churchill Falls, Newfoundland, is rated at over 5 million kilowatts, and will supply power for Quebec and Montreal, 700 miles away. An even larger installation, rated at 14 million kw, is being planned for James Bay, some 700 miles to the west. While some of this giant capacity is to be used to fill existing needs, and some to spur development in the area, some part may also be sold to the United States to fill the rapidly growing needs there.

In the United States, while we will see further development of hydroelectric power, it is likely to constitute a falling percentage of the total. Today, as we have seen, it makes up about 16 percent of our total generating capacity. The Federal Power Commission forecasts that this will drop to 12 percent in 1990. (The National Petroleum Council sees it falling to 7 percent by 1985.) The F.P.C. also calculates an ultimate maximum of about 161 million kilowatts in the United States, and a world capacity of 2.8 billion kilowatts.

Roughly speaking, total potential hydropower is about equal to total electrical requirements worldwide. But this sounds more hopeful than it is. Most of the potential is in presently undeveloped areas. There are also two major objections to hydropower.

The first has to do with the fact that some of the world's most sensational scenery may have to be sacrificed to this kind of progress. Typical are the steep, rugged gorges that have been cut into mountainous areas by rapidly flowing water. Building a dam means that the gorge becomes filled with water. These provide a natural wall on the sides for the reservoir that builds up behind the dam. For a large installation, thousands of acres of beautiful scenery may be lost.

There is another problem. Rivers normally pick up sediment in the course of their travels. Some of it may be deposited on adjacent "flood plains," thus fertilizing them; but most often the sediment is carried out to sea. If a dam is built, the water slows down and drops these sediments behind the dam. Eventually, perhaps in as little as a century, the reservoir fills up with sediment. While the stream flow may continue, and the dam then acts like a waterfall, there is no longer the control offered by the dam and the period of maximum usefulness is past. Perhaps, however, by that time engineers will be capable of dredging out the enormous amounts of sediment.

Large amounts of good land may also be drowned. Being near a river, it is likely to be good farmland, particularly in the more humid eastern sections of the country. Wildlife habitats may be disturbed or destroyed, and fish injured or killed. Leakage from the reservoirs may raise the water table in nearby areas, which can raise the likelihood of flooding. The very weight of the water can cause such problems as earthquakes. And, of course, damming up the water creates a potential hazard if the dam should break, as has happened several times in recent years.

All in all, while it is certain that hydropower will continue to be developed, it is not likely to prove a mainstay of our energy resource base, particularly since we are seeing a greater interest in preserving our environment; already more stringent laws to preserve rivers in their natural wild state are being or have been passed.

Tidal Power

Among the more impressive ways of generating power that man has tried is putting the world's tides to work. As the earth, moon, and sun change their relative positions, the changing gravitational pulls of these two bodies on the oceans of the earth cause a twice-daily rise and fall in the tides. Various configurations of bays and riverbeds can cause the difference between low and high tide to be as much as 55 feet, though this is most unusual. As with conventional hydroplants, the difference between high and low water can be exploited to produce electricity. In this case water is allowed to come through a gate into an enclosed area at high tide. When the tide begins to ebb, the gate is closed, trapping the water there. At low tide another reservoir, the low pool, is emptied. Then all the gates are closed and the water-level differential can be held in readiness for a peak demand time. If the peak time happened to coincide with low tide, the low pool wouldn't be necessary.

In the United States a number of promising sites exist, varying in potential output from perhaps 2000 to 20 million kilowatts each. The best known is the Passamaquoddy Tidal Power Project off the Maine coast which, with a potential output of 300,000 kw, has been in the discussion stages since the 1920's. The tidal range here is about 18 feet. The high capital cost of development, plus the readily available alternative supplies, have prevented actual construction so far.

The only large-scale tidal project in actual operation is one on the Rance River in France. Its capacity at start-up in 1966–67 was 240,000 kw; final production of power is expected to go to 320,000 kw.

M. King Hubbert estimates that the total potential tidal power of the earth comes to about 64 million kilowatts, or only about 2 percent of the world's potential water power.

Clearly, tidal power can never become a major contributor to the world's energy budget. In those few areas lucky enough to have the proper conditions, however, it can provide convenient, pollution-free power, and without the environmental demands of conventional hydropower. (An allied system, pumped hydrostorage, is described in Chapter 13.)

Power from Waves

Any mechanical motion that is relatively steady can be used to power a generator. A common example is the hand-powered flashlight that provides light as long as the handle mechanism is being "pumped."

Consider the following quote from an early science fiction book, *We,* first published in English * by the Russian writer Eugene Zemiatin:

". . . in the days of the ancients [that means us] the ocean blindly splashed on the shore for twenty-four hours a day, without interruption or use. The millions of kilogram meters of energy which were hidden in the waves were used only for the stimulation of sweethearts! We obtained electricity from the amorous whisper of the waves! We made a domestic animal out of the sparkling, foaming, rabid one!"

Well, that's one approach to an ever-renewable energy supply. Most technologists would say it's not a very hopeful source, at least not for central station power, sharing with the wind a problem of unevenness. But I surely won't say it can't be done.

In Japan the motion of waves has been put to work, however, to power the lights and foghorns of harbor buoys. These buoys are reportedly simpler to service than those using conventional or even solar batteries. Two types have been built: One uses the vertical movement of the waves by means of a plunger mechanism; the other is a pendulum type that

* E. P. Dutton & Co., 1924.

utilizes the rocking motion imparted by the water. While we can hardly count on such devices for any large-scale power needs, ingenious applications of natural power from the earth can reduce requirements and thus help in conserving our resources.

For large-scale power needs, a very different approach is necessary. In the next chapter we take a look at what is probably our greatest hope for the future.

11

Fusion

In fission, heavy elements such as uranium are made to split and release energy in the process. In fusion, another nuclear process, the reverse takes place: Light elements, such as hydrogen, fuse together into heavier ones. And in this process, too, more energy is released once it is gotten going, than it takes to get it started.

The problem lies in getting it started. Fission is easy by comparison—which is why we already have fission plants while scientists are not even certain, after 20 years of research, that the fusion process can be made to work in a controlled fashion here on earth. That fusion can be accomplished here on earth has already been demonstrated in the uncontrolled fury of the hydrogen bomb. And that it can work in a controlled sense is known from the fact that this is the process which provides the heat and light of our sun. (By controlled we mean that it continues at a steady rate and does not "run away," or build up into an explosion.)

Fusion fuels are, in general, common, accessible, and

cheap; and they "burn" cleanly, which is not true of fission fuels. Fusion also promises high efficiencies—some say up to 85 or 90 percent in certain configurations—and greater potential safety than fission.

Strangely enough, the scientists who first thought of extracting energy from the atom visualized fusion as the route that would be taken. But fission turned out to be a much easier conquest. The later demonstration of the awesome power of the hydrogen bomb once again set scientists to thinking about controlling it—putting it to constructive work.

For fusion to occur, two light nuclei must smash together at high enough speeds (about 22 million miles an hour) to cause them to fuse. In gases the velocity of movement, the energy, and the temperature of particles are all mixed up together. As a result, one way to get gas particles to fuse is to heat them up to extremely high temperatures—tens of millions of degrees. Thus another term for fusion is controlled thermonuclear reaction, or CTR for short.

In the sun the process takes place at 25,000,000° F. and very high pressures. The raw materials are hydrogen nuclei; these are hydrogen atoms that have been ionized, or stripped of their electrons. At such temperatures all gases are ionized, leaving a group of charged particles (mainly protons and electrons) that, if allowed to come together, would make up an electrically neutral gas. Such a gas, as we have seen, is called a plasma.

Containment

We have already noted how the responsiveness of plasma to electromagnetic forces has been applied in MHD (direct conversion). This characteristic is being put to work in the CTR field as well, namely, to confine the plasma to a particular working volume. A physical container cannot be used —at those temperatures any substance would vaporize and, more important, the reaction would be quenched immediately.

So the approach has been to use magnetic containment. An extremely strong magnetic field is set up in such a way that a kind of magnetic bottle is created. This approach can work because, although the over-all gas is neutral, the particles are charged and respond to electric and magnetic fields. (On the sun the process is operating on a tremendous scale, so that the gravitational force does the work of confinement.)

The objective is to hold the gas together long enough to bring it up to the temperature and pressure required to initiate the fusion process. Although it does take a large amount of energy to bring the temperatures and pressures up, much of the input energy in present methods goes into the magnetic fields used for confinement (the advent of superconducting magnets has been of great help here), for cooling, and for other extraneous requirements.

But the main difficulty is that artificially confined high-temperature plasmas are extremely unstable. They seem almost alive in their attempts to bend, squirm, skitter, or otherwise escape the magnetic vise set up to contain them. The straight hydrogen-to-helium route that takes place on the sun is too difficult to get started and too slow for economical use on earth. It works on the sun because of the enormous amount of fuel that is present there, as well as the great densities involved. Fortunately, hydrogen comes in several varieties (isotopes), which can ease somewhat the process of getting the process going.

A normal hydrogen nucleus consists only of a proton. Hydrogen with 1 neutron added becomes deuterium; and with 2 neutrons becomes tritium. Deuterium (or heavy hydrogen)

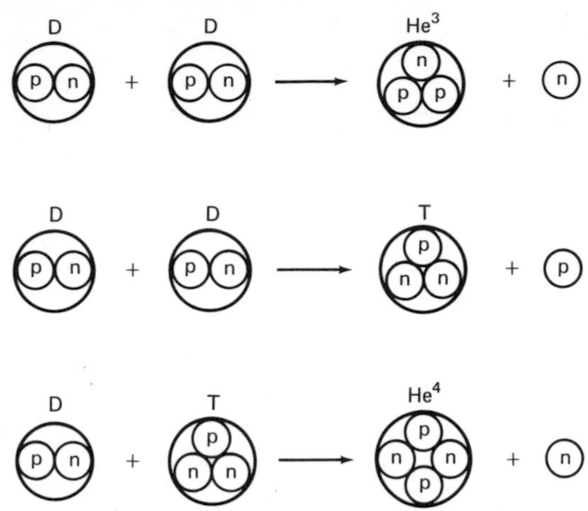

Shown here are the fusion reactions of greatest interest at present, those involving deuterium (D) and tritium (T). The two D-D reactions are equally probable; the D-T reaction is of interest mainly because the ignition temperature of 40,000,000° C. is the lowest known on earth. If it is found that the D-T reaction is the one we have to resort to, the tritium would have to be "manufactured" by bombarding lithium-6 with neutrons (n): $^{6}Li + n \longrightarrow T + {}^{4}He$

At first the neutrons would probably have to be gotten from a fission reactor. But after the fusion process was begun, the neutrons released in the D-T reaction could be used to continue the process.

is found in all water to the extent of 1 part in every 6700 of ordinary hydrogen. While this does not sound like much, the quantities of water are so extensive that the supply of deuterium (D) is virtually inexhaustible. One calculation says there is enough to last a billion years, even at increased usage rates. And the cost of separating out the D is negligible compared with the energy that can be released from it. Each gallon of ordinary water contains deuterium with an energy value equivalent to 300 gallons of gasoline.

Tritium (T), on the other hand, exists only in very tiny

amounts in nature. And so while a D-T reaction (see diagram) would take place at a much lower temperature than a D-D reaction, there is far less such fuel. At this point, however, the difference is academic, since we have not been able to accomplish a continuous, controlled reaction with either one.

We are coming close, however. And it is only in recent years that it has been possible to say this in real honesty.

Requirements

Indeed it looks for the first time as if we are approaching the point where scientific feasibility of CTR will actually be demonstrated. This would be more or less equivalent to the atomic pile experiment at Stagg Field in 1942, or the Kitty Hawk flight by the Wright Brothers in 1903.

The definition of scientific feasibility in fusion might be production of a plasma that produces a sustained fusion reaction which yields more energy than has been put in to get it going. Three basic conditions are involved: 1) the density of the plasma, or the number of particles (n) per unit volume, 2) the time of containment (τ), and 3) the temperature. The desired combination is generally described by the Lawson criterion, which calls for $n\tau = 10^{14}$ sec/cm^3 at 50,000,000 to 100,000,000° centigrade. Thus if we could maintain 10^{14} particles per cubic cm at that temperature for one second, or 10^{15} particles * for $1/10$ of a second, this would be considered a successful demonstration of feasibility. Each of the various criteria has been attained individually, but never all together.

* This is approximately one ten-thousandth the density of the atmosphere at sea level.

Due to the great complexity of the behavior of hot plasmas in magnetic fields, a large number of approaches are being investigated. Basically the magnetic confinement schemes fall into open and closed systems.

A closed system might be shaped like a doughnut, with the plasma circulating through the hollow interior. The advantage is that there is no end out of which the plasma can escape. The fact that the ring is curved, however, means that the field is not an even one, being more concentrated at the inner portion, which leads to greater difficulty in maintaining an even confinement. The international favorites are two devices called the stellarator and the tokamak.

It is in the latter device that the hopeful developments mentioned earlier took place. This was accomplished by Soviet scientists in the years 1968–69. For one thing, they came closest to the Lawson criterion, attaining an $n\tau$ of 10^{12}, though at a temperature of less than 10,000,000° C. But more important, they showed that as temperature is raised, the confinement-time situation improved, as theory said it should. Prior to then, each rise in temperature seemed to make the situation worse. In other words, a corner was turned. Oddly enough, though the Russians had made this important step, they didn't have the equipment to prove it, while the British did. In a fine demonstration of international cooperation, a team of British scientists took their equipment to Moscow in 1969 and made the necessary measurements of the velocities of the gas, bearing out the Russians' claim.

The news electrified the scientific world. Very quickly thereafter a number of American and foreign machines were being adapted to this configuration, whose major advantage, oddly enough, is its simplicity. It is this characteristic that enables it, in both an economic and technical sense, to be made larger. Its disadvantage is that it is less flexible and

A tokamak at work.

harder to use in experiments that require measurements.

Shown here is one of the larger such machines, located at the Princeton University Plasma Physics Laboratory. Princeton has just been given a go-ahead to build a considerably larger tokamak, the PLT (Princeton Large Torus). While this will still not be large enough to bring us up to the Lawson criterion (we are still shy by a factor of about a thousand), it will show whether this is the right direction to take.

In essence, what is done in such machines is to pass a large current through the plasma contained in the doughnut or torus, thus heating it, as in the resistive coils of a toaster. At the same time the plasma must be held together and prevented from squirming out of shape and thus touching the container walls; this is done by maintaining or increasing the strength of shaped magnetic fields.

Heating

It is known, however, that resistive heating will not continue to work at still higher temperatures, for the resistivity of the plasma begins to drop. At this point other approaches become necessary. Among them are increasing the turbulence of the plasma, injection of neutral, high-energy particles,

The author peering into one of the toroidal field coils of the Princeton tokamak, which had been disassembled for servicing.

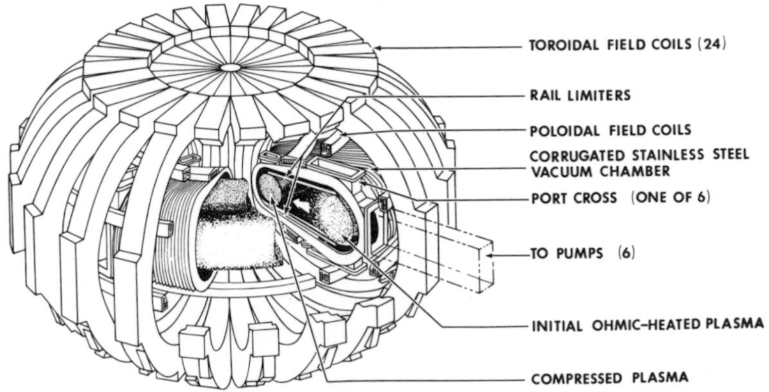

Adiabatic Toroidal Compressor.

and compression. The Texas Turbulent Tokamak (University of Texas) speaks for itself, and the ORMAK (Oak Ridge Tokamak) will perform experiments with neutral-particle injection. The compression approach is represented by a Princeton machine, the ATC (Adiabatic Toroidal Compressor) which is shown in cutaway.* What happens is that the plasma ring is heated as high as is practicable by resistive heating around the periphery of the machine. It is then magnetically forced in toward the center, which increases the pressure and temperature proportionately.

Another major approach to confinement lies in the "open" machines. These do not close on themselves, as do the ring types. While containment becomes less complex in the sense of being a more even problem (no curves), there is then the problem of leakage at the ends of the plasma "bottle." In this case, what is attempted is to increase the field strength

* If there seems to be a heavy emphasis on Princeton machines, it is only because the author visited that laboratory—and came away much impressed. Actually, important work is going on in a number of labs.

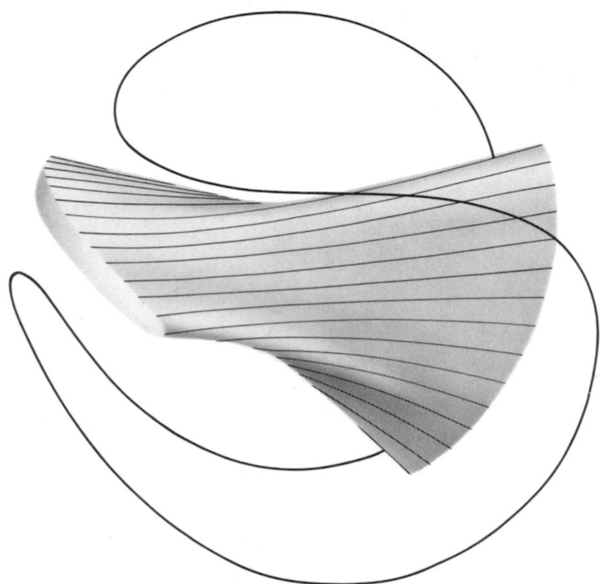

The windings in a "baseball" machine, so called because of the baseball-seam-like windings that mirror plasmas for fusion.

at the ends of the bottle, thus creating a sort of magnetic mirror. Presently, three new magnetic mirror devices are beginning experiments in the United States—Baseball II and 2XII, and IMP.

In the IMP experiment, taking place at Oak Ridge National Laboratory, the objective is to produce plasma heating by injection of microwave energy, rather like that done in microwave ovens. IMP thus stands for Injected Microwave-heated Plasma.

The Baseball machines are so called because their field windings are in the strange and interesting shape of a baseball seam, which leads to the type of plasma shown. The idea is to create a kind of magnetic well, though the problems with leakage at the ends are obvious. But this may be taken care of by an interesting suggestion made by Dr. Richard F. Post of the Lawrence Radiation Laboratory, at Livermore, California. He feels that we may be able to turn a problem into an asset by using this leakage. After all, he says, the leakage is of charged particles. Why not then increase this leakage to the point where it can be used to power an MHD generator, thus producing a direct conversion cycle? Such a combination might bring forth the very high efficiencies mentioned earlier.

Although we have by no means exhausted the list of magnetic-confinement concepts (it may very well be one we have not described which will eventually make the breakthrough), we must move on to a rather different approach.

Proposed fusion reactor with direct energy converter.

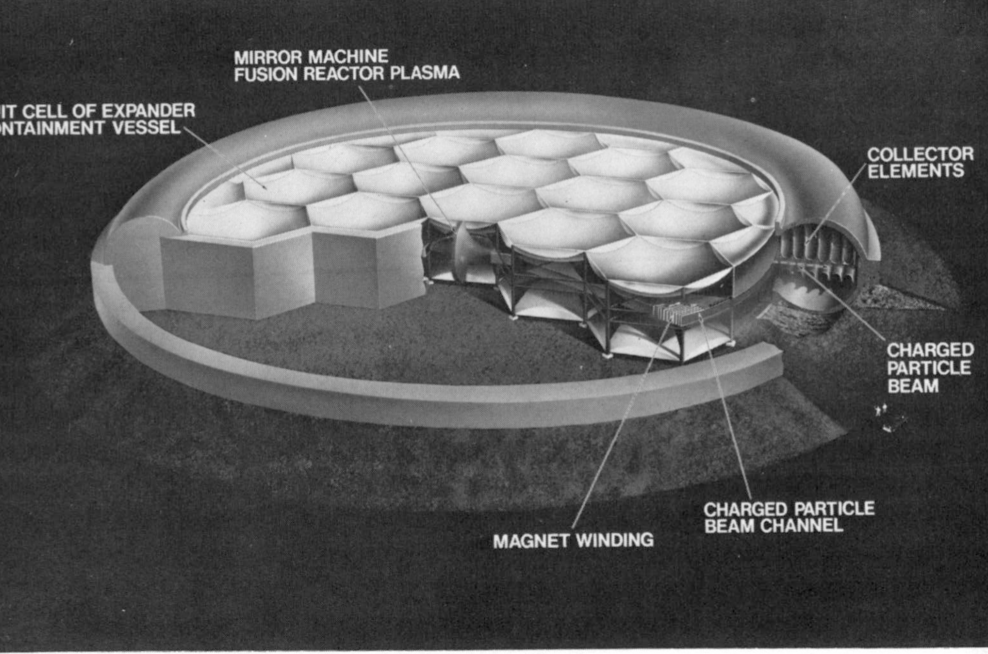

Light Beams and Electron Beams

The rapid development of high-power, pulsed lasers during the past decade has brought forth a suggestion that such a device might be able to provide the energy needed for heating a fusion plasma. The main point is that if a "chunk" of matter can be heated quickly enough, it might not be necessary to contain the resulting plasma at all. Its own inertia might hold it together long enough for the fusion process to take place. Remember that the Lawson criterion had mixed in together both density and containment time. If the density is high, only a very short time is needed. In this case it would be the time it takes for the reacting substance to fly apart.

As Dr. Roy W. Gould, until recently assistant director for the CTR research program of the Atomic Energy Commission, explains it, "This approach to fusion is characterized by very high energy densities for very short times—in essence, a microexplosion." A power plant using the laser approach might involve the production of 10 to 100 such microexplosions per second, which is roughly comparable to the rate at which explosions take place in the internal combustion engine of a car. Interestingly, it was again a Russian who, in 1968, announced the first observation of neutrons produced by laser heating.

Although lasers have increased enormously in power during the past decade, they are still 1000 times too low in strength. A further complication is that it is not sufficient to "hit" the pellet from one side. It must be hit from all sides at once if any kind of containment (in essence, an implosion) is to take place. This is a complicated enough problem in its own right, though perhaps solvable by optical beam-splitting means. But how do we then surround the pellet with a material (a "blanket") that permits us to bring out the energy? Dr. Edward Teller, one of the early workers in the thermonuclear field, expects these problems to be tough ones to solve, and suggests that we may see something else sooner: The use of this sort of device as an actual internal combustion engine to power rockets! Nevertheless, in May of 1972 the University of Rochester teamed up with the General Electric Company and Esso Research and Engineering Company in a $3 million program to work on the laser and fusion.

In a somewhat similar program of research into the problem, beams of high-energy electrons are being tried as the source of heat energy.

And, finally, it has been found possible to store microwaves in special containers with superconducting walls at intensi-

Laser beams striking a tiny deuterium-tritium pellet would cause the outer portion to blow off and the core's atoms to fuse together (implode), creating heat.

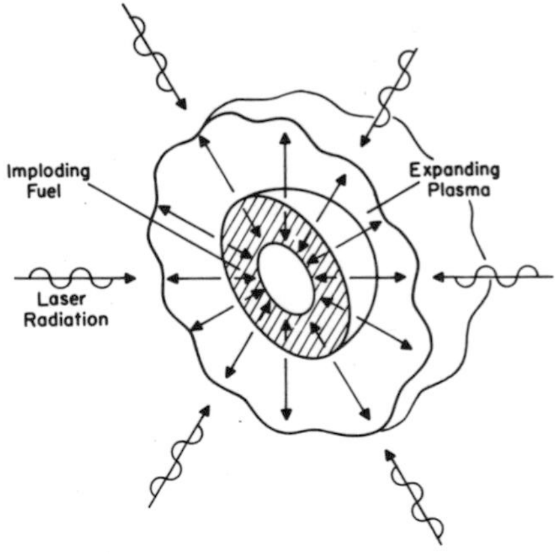

ties a billion times higher than that of sunlight, high enough to explode instantly any ordinary or unprotected substance with which it comes in contact. Perhaps such stored energy can somehow be used in CTR work.

Reactor Designs

Which, if any, of these approaches will win the fusion derby is impossible to say at this time. Most of the effort in fusion research (some $500 million to date in the United States alone) continues to go into understanding the plasma itself. It is a measure of the rapid movement in the field that we are starting to get calls from researchers for investigation into actual reactor configurations. But when we start to look into reactor designs, other problems pop up, such as how to extract the resulting energy. This will depend to a large extent on which reaction is used. In the D-T reaction, 80 percent of the energy is released in the form of highly energetic neutrons. Present expectations are to absorb the neutron energy in a liquid lithium shield or blanket perhaps a meter thick. The lithium will be continually circulated to a heat exchanger, as in the LMFBR, where it will produce steam for use in a conventional steam-generating plant. At the same time, tritium will be bred in the blanket and then extracted for further use as a fuel.

It may also be possible to use some of this abundant production of neutrons to breed fuel for fission plants! This could increase the world reserves of thorium-232 and uranium-238, and at the same time would improve the economics of the fusion aspect, thus bringing it to a point of commercial practicality sooner than otherwise would be the case.

The advantage of a D-D reactor is that it would not be necessary to breed tritium (which is rare in nature) by irradiation of lithium (in the blanket). A simpler water blanket could be substituted, though this would mean operating at higher pressures to maintain a high-temperature regime. Also, about one-quarter of the energy output could be taken out directly in the form of electricity.

Prospects

It is expected that practical, economical fusion plants will have to be approached in step-wise fashion. First must come demonstration of scientific feasibility. The complications that still lie ahead are shown by the fact that in spite of the new enthusiasm and optimism, few expect this to occur much before 1980.

After feasibility is demonstrated, the successful approach or approaches will then have to be duplicated in demonstration or prototype plants that will be far larger and more expensive than the present class of research devices. It is characteristic of most of the fusion approaches that they become intrinsically more efficient, and even possible, as they get larger.

As with breeders, it will be necessary to build at least two or three demonstration plants to be sure that, as the size begins to approach reactor size, other problems are not creeping in. Representative Craig Hosmer of California, a member of the influential Joint Committee on Atomic Energy, maintains that the development of a fusion reactor is perhaps 50 times more complicated and difficult than that of a fission plant, which took some 25 years from demonstration to large-scale, economical plants. The greater difficulty is somewhat counterbalanced by the greater amount of experience we have developed in the nuclear field.

Dr. Edward C. Creutz, assistant director for research at

the National Science Foundation, foresees some initial sales by 1990. Others are more pessimistic. The consensus seems to be that we will not see fusion-reactor plants until the beginning of the twenty-first century.

Much will depend on how much support is given fusion research in coming years. Current support runs some \$30-\$40 million per year, most of it from the government by way of the A.E.C.

But this will have to go up substantially as we start to go into the larger machines. (Breeders are currently being supported by the A.E.C. at a rate of about \$100 million a year, and by industry as well.) The U.S.S.R. seems to be spending at about twice our rate on fusion work. Most of the American work is going on at the four major laboratories—Princeton, Oak Ridge, Los Alamos, and Lawrence Radiation Laboratory, with a smaller amount going on at university and industrial labs.

The final result, hopefully, will be plants that can supply us with an inexhaustible source of power with minimum impact on the environment. It is claimed that they will be so safe that they can be located right inside urban areas. This would eliminate the need for transmission of power from outlying areas, and would also make it possible to use the excess heat for space heating, distillation of sewage, etc., that normally must be dissipated into great quantities of cooling water.

Although the supply of power will be virtually inexhaustible, it will not necessarily be cheap. One estimate puts the cost at the same level as breeder reaction power. Yet a saving of even 1 mil (1/10 of a cent) per kilowatt hour could mean a saving of \$10 billion a year by the next century, so huge will the electrical usage rate be at that time. In other words, the savings in one year might well be more than the total cost of development, which at this point looks to be some $5–8 billion.

CTR is not a panacea, for there will be some concentration of tritium, a radioactive material. The advantage that fusion reactors have over fission devices in this respect is that the tritium can be recycled right at the plant, thus eliminating one major source of worry today, shipment of radioactive materials. Another major advantage is that there will be no radioactive waste materials that must be stored.

It seems that one would have to agree with the Electric Research Council, which stated in one of its reports, "There is no question that the achievement of a practical fusion power reactor would ultimately have a profound impact on almost every aspect of human society."

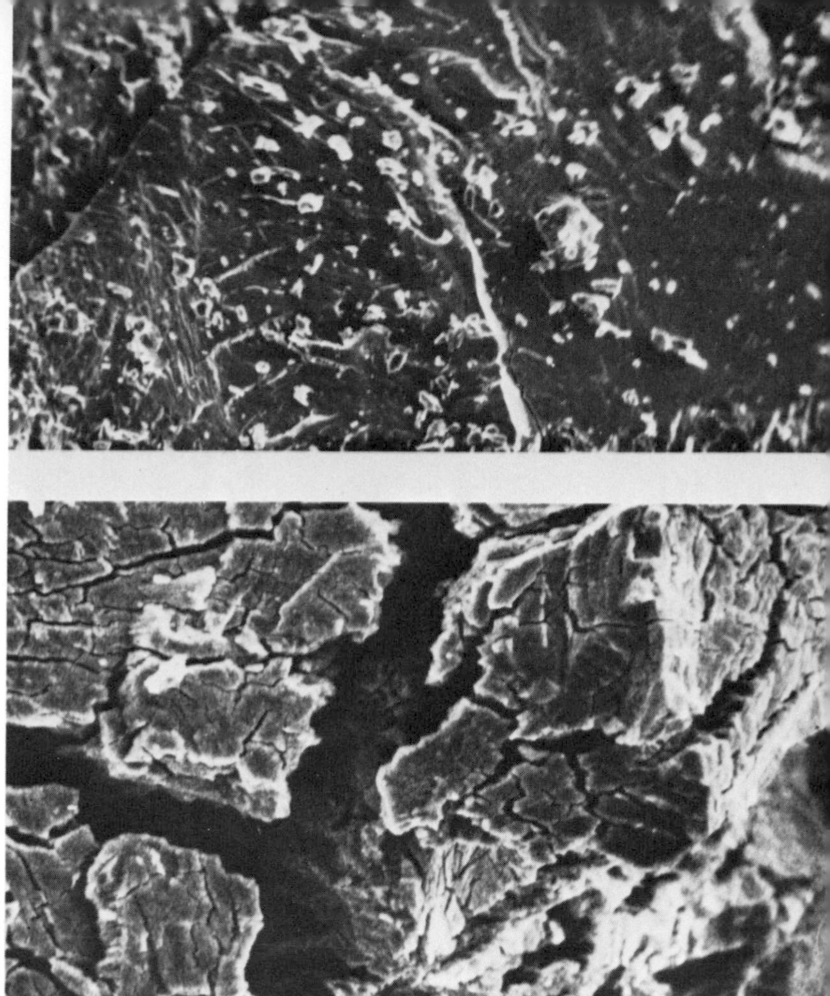

A look inside your "gas" tank of the future? To get rid of the pollutants all fossil fueled engines now spew into our environment, your car of the future may use hydrogen, which is almost totally free of pollutants, as a fuel. (See page 154 for a discussion of hydrogen fuel.) Your "gas" tank will appear to the naked eye as solid metal titanium, or magnesium alloy granules, instead of gasoline. Shown "empty" (without hydrogen) in the top photo, but "filled" (hydrided) in the bottom photo is an iron-titanium alloy taken at a magnification of 500 times. (Brookhaven National Laboratory.)

12

Some Speculative Energy Schemes

WE HAVE SEEN two levels at which energy can be obtained from matter: combustion and nuclear processes. In the first, matter is not being consumed but merely rearranged. And even in nuclear energy only a tiny portion of the mass involved is actually converted to energy—less than 1 percent. But, as we know from Einstein's familiar formula, $E = mc^2$ (where E is energy, m is mass, and c is the speed of light), the conversion of even a little matter can supply quite a bit of energy.

This suggests yet a third level of energy production—the complete conversion of matter to energy. The huge energies apparently being released by the recently discovered quasars lead to speculations that some as yet untried energy production mechanism (untried here on earth) is at work. One possibility is the total conversion of mass to energy through the meeting of matter and antimatter. It is believed that each elementary particle (proton, neutron, etc.) has a counterpart (antiproton, antineutron, etc.). Antimatter is known to exist, having been seen in the debris of collisions of cosmic rays with particles in test devices, and also having been produced in particle accelerators.

When separate, a charged particle and antiparticle appear to be different only in having an opposite charge. But if they come together, they disappear in a puff of, not smoke, but energy. A meeting of an electron and its opposite (a positron) will give off over 1 Mev* of energy in the form of electromagnetic radiation (e.g., gamma and X-rays). A proton and its opposite number will release some 1900 Mev. By comparison, the three fusion reactions shown on p. 132 would release, respectively, 3.2, 4, and 17.6 Mev.

Soviet scientists have reported evidence of antiparticles in space, which would tend to support a widely held hypothesis that the universe contains antimatter galaxies as well as "normal" ones like ours. They admit, however, that what they saw might have come from cosmic ray collisions with normal particles in space. Perhaps the most valuable cargo some future space expedition will bring back from a neighboring galaxy will be a couple of pounds of antimatter.

How one would package such "fuel" and derive useful energy from it is a question we can't get into here, except to suggest that the use of electromagnetic fields, as in MHD, would probably find use.

While we're on the subject of particles, we might mention another strange possibility. A few years ago it was suggested that there might be particles which, like photons, have no existence except while traveling. But photons, the particle car-

* 1 Mev is 1 million electron volts. One electron volt (ev) is the energy developed by a unit charge, such as an electron, in falling through a potential difference of 1 volt. 1 Mev $= 15{,}187 \times 10^{-17}$ Btu. See Appendix for conversion factors.

riers of light, travel *at* the speed of light; the new particles, called tachyons (from the Greek word for fast), can travel at speeds only faster than light!* This, it is suggested, is one reason no one has seen them yet.

The idea is highly speculative at present, though some quasars at least appear to be expanding faster than the speed of light. If such particles exist, they would have some strange properties. One is negative energy, which is a little difficult to visualize. A bank account can have a negative balance, but what does one do with the idea that someone weighs minus 140 pounds? Let us imagine however, that a tachyon is emitted by a proton (as a particle might emit a photon). Then the loss of negative energy caused by emission of the tachyon should result in a gain in the positive or normal energy of the emitting particle. This might show up in a sudden increase in its speed. Such an effect could surely be put to work in the energy field.

The potential of tachyons numbs the mind. How about messages traveling at 10 or 1 million times the speed of light? Or, while we're at it, space travel that doesn't require 4⅓ years to get to the nearest star, or several dozens of years to get to a certain galaxy (to get some of that antimatter). Experiments have been performed, looking for tachyons, but so far in vain.

Central Station Power

The technological dream of safe, clean, unlimited, and cheap energy can come in one or both of two styles. The first

* Proponents of the idea argue that while Einstein's special theory of relativity specifically excludes accelerating objects to speeds faster than light, it says nothing about particles that have never been at lower speeds.

might be a decentralized kind of power system, similar to a self-contained diesel generator set that is used to provide power today for some out-of-the-way military installation. One can imagine, however, that the device has been brought down in size to that of a fist or a gallon can, so that it can be used to power a car or boat or, by putting a few together, a house, train, or plane. The second style would be a better way than we now have of providing central station power which is in turn distributed. The possibilities we have just discussed could fall into either category, but would most likely fall into the first.

One rather far out possibility as a source of central station power is the potential energy of rain, or even the latent heat released when water vapor changes to rain. It takes energy to convert water to water vapor. Similarly, energy is released when water vapor changes back to water. The energy released, though spread out, is enormous. One inch of rain falling over a 100-square-mile area during a storm would release more than 10^{13} Btus, or the energy equivalent of 175 atom bombs of the World War II type. Most of this heat energy would be converted to energy of motion at a very low efficiency—about 1 percent—but nevertheless represents, at 2 cents per kilowatt hour, about $1 million.

Or consider this idea: We build a dome over a city. Not only could the entire city then be climate-conditioned, but all rain and snow that fell on it could be collected for fresh water. The dome could be built of material that darkens in hot sun, thus making more efficient the collection and application of solar energy on a large scale. Special electrodes might somehow tap the potential difference in electric field that exists between sea level and upper altitudes, particularly during electrical storm periods. Even lightning itself might be put to work; the average lightning stroke contains an energy of about 1000 kw-hr. Other collectors might be able to tap the energy of incoming electromagnetic radiation and particles from the sun and space. It seems likely that wind power, and even the incalculable energy of major storms, might somehow also be harnessed.

Or how about the rotational energy of the earth and/or moon? We might think of the turning earth as the permanently magnetized rotor of a huge motor. Could we somehow put to use some of the 245×10^{24} Btus of energy contained therein?

The highly respected Princeton physicist Freeman Dyson suggests that if a society's technological prowess continues to grow at a steady rate, in a few hundred years its control of mass and power should increase by factors of millions or even billions, Eventually, he says, it must have complete control over the resources not only of its own planet, but of its entire solar system. It might eventually need so much energy that it would have to enclose the sun and tap its energy directly instead of going through the various inefficient conversions such as we use here on earth!

"Growing" Fuel

One such conversion that is often suggested as a potential fuel source is photosynthesis. We have seen that fossil fuels derive from decayed living matter, which got its energy, directly or indirectly from photosynthesis. To put the process to work on a "here-and-now" basis means, at least at present, growing substances that are later to be converted somehow into fuels (e.g., growing corn which, by fermentation could yield burnable alcohol).

This can be, and has been, done. During World War II, when there was a worldwide shortage of oil and gasoline, plants were built in India, Pakistan, Brazil, and Jamaica to convert plant substances into alcohol, which is a usable material for powering cars and trucks.

But this method of "growing" fuel is inefficient, and its cost would be many times that of the equivalent in oil. Another objection is that serious problems could arise if large investments were made in this method only to find later on that the land devoted to it became needed for food or other purposes. With ever-increasing populations, even here in the United States, this is very likely to be the case.

The process of photosynthesis is extremely complicated. It is not completely understood, and probably won't be for another decade or so. Two men who are studying the problem are Dr. James R. Bolton of the University of Western Ontario and James T. Warden of the University of Minnesota. In a paper presented at a meeting of the American Chemical Society,* they stated:

> "The primary photochemistry of photosynthesis can be linked to a little photo cell which converts the light energy directly into electrical energy. The plan then uses this electrical energy to drive various chemical reactions. It may be possible to convert light energy into electrical energy at very high efficiencies. Such a system of power production would not only be highly efficient but also would be totally pollution free."

An interesting alternative has been suggested by Arthur E. Sowers of Texas A & M University. He writes in a letter to *Science News*: *

> "Suppose we learn enough biochemistry and molecular genetics to breed, mutate, and otherwise genetically alter photosynthetic algae to 1) 'over-produce' one of their cell

* August 30, 1972.
* June 17, 1972, p. 386.

membrane constituents (fatty acids) and 2) decarboxylate them to 'oil' (long chain alkenes). These oils would pass to the cell's exterior and, being insoluble in, but less dense than, water, float to the surface for convenient collection. . . . One can show that a square of about 100 miles on each side would be needed. This is large, but still less than Lake Michigan. In any case, it would be a reasonable price to pay to get an unlimited source of oil. . . . It might be possible to make it capable of self-regulation and self-repair, like many naturally occurring ecosystems."

Hydrogen

While all of the various energy alternatives we have discussed have been, or are being, looked into, the main bulk of actual research in the energy field has gone and continues to go into nuclear energy. This is rather strange when you think about it. For except for some small use of portable power devices, used mainly in space and medicine, it is all central station power. Yet, at the moment, this makes up only a quarter of our total energy needs, though this is expected to increase. And nuclear-generated electricity cannot under present circumstances act as a suitable substitute for our dwindling supplies of oil and gas—at least not without vast changes in the transportation field as well as in residential, industrial, and commercial areas.

The most pressing problem is that of the growing shortage of fossil fuels. In order not to completely overturn our present transportation system, which all-electric vehicles would probably do, we must somehow find a non-fossil chemical fuel. It has already been seen that cars, trucks, and buses can run quite satisfactorily on natural gas and similar fuels. The operator of a taxi fleet in Red Bank, New Jersey, has recently converted all his cabs to run on propane fuel (a gas at atmospheric pressure that can be liquefied, for convenience in storage, under pressure). It burns cleanly and eases maintenance problems. He says operating costs have dropped 32 percent.

Such vehicles can also be run on hydrogen, which burns even more cleanly than natural gas; the only emission is water vapor and a small amount of nitric oxides (about 1/10 that of the present internal combustion engine).

The main advantage to using hydrogen is that it is not in limited supply, as are fossil fuels. It is present in the atmosphere and is a major constituent of water. Present schemes foresee splitting water into hydrogen and oxygen, which is already being done for other purposes, but on a huge scale. Both gases are already finding use in various parts of the economy. Hydrogen is used for hydrogenation of fats and oils (addition of hydrogen) in making ammonia and other chemicals, and as a reducing agent in the manufacture of certain metals. But it wasn't until the gas began to be used on a fairly large scale as a fuel by the space people that its price came down considerably and made it begin to look like a potentially economic and useful fuel.

Electricity can be used as the primary energy input in production of the gas (by the process of electrolysis). It would most likely be obtained from nuclear power plants. Heat, which is also produced in abundance by nuclear plants, could be used too, and therefore also suggests the use of solar energy. Another primary source might be ultraviolet rays, which can be produced in copious quantities by fusion devices.

If hydrogen were introduced on a wide scale, the presently installed pipeline systems, storage tanks, and production plants now used for natural gas could be used immediately for hydrogen. Everything that natural gas can do, hydrogen can do too. The same holds for coal and oil. For some applications, e.g., as a motor fuel, it may be necessary to compress or liquefy it. The gas boils at about −253° C., or −423° F., and so must be brought down to this cryogenic temperature for liquefaction. This adds some complication and cost but is not an insurmountable obstacle.

An interesting alternative arises due to the fact that hydrogen is rapidly absorbed by certain metals, such as magnesium. It could thus be stored in a porous magnesium tank. A 500-gallon tank could hold the energy equivalent of 20 gallons of gasoline—which means you could leave your car for a couple of weeks without the fuel disappearing, as it would because of boil-off if left in a liquid state. The heat

The automobile engine of the future? A standard, 3-horsepower gasoline engine modified to burn gaseous hydrogen. The injection mechanism is at upper left.

of operation of the auto would liberate the fuel as needed. (A 500-gallon tank, however, would occupy 67 cubic feet, or a cube with sides four feet across!)

An alternative would be to take the hydrogen from water and combine it with nitrogen from air to produce either ammonia, which is also a burnable gas but one that is more easily stored than hydrogen, or hydrazine, which is liquid at atmospheric temperature and pressure.

The light weight of hydrogen coupled with its high energy value give it the highest energy density per pound of all fuels. This would be particularly useful in air flight. A hypersonic aircraft capable of flight at more than 3000 miles per hour has already been proposed. Not only would the supercooled liquid hydrogen be used to fuel the plane, but it would also provide an answer to one of the major problems of high-speed flight through the atmosphere: heat. Temperatures on leading-edge surfaces can go up into the thousands of degrees, causing weakening of structural members. The supercooled liquid hydrogen, at −423° F., can be used to cool a secondary fluid that would in turn be circulated through the friction-heated areas.

There are some dangers as well as inconveniences connected with the use of this unique substance, as anyone who has heard about the fate of the hydrogen-inflated dirigible, *Hindenburg,* will immediately recognize. The two main threats are fire and explosion. But those who espouse this new use maintain that gasoline too is a highly dangerous fuel, yet no one seems to worry too much about it anymore. In addition, tank cars of compressed or liquefied hydrogen pass through our cities every day with no notice being taken of them at all.

The prospect is for clean, flexible, practically unlimited fuel, with relatively few disadvantages. Will we see this come to pass? Perhaps, but probably not until things get considerably worse in the oil and gas industries. With the billions of dollars that large oil, auto, and other companies have invested in the present system, they are not likely to make such a change voluntarily. Nor are we the public likely to give up easily the admittedly greater convenience of the gasoline engine. In addition, each and every one of the millions of furnaces in the country would have to be changed, rebuilt, or adjusted.

All in all, the problems associated with a changeover to a hydrogen economy are immense, though they seem at present to be strictly technical in nature; no scientific breakthroughs are needed.

The automobiles of the next 5 or 10 years may be able to meet the increasingly stringent requirements of the Clean Air Acts of 1967 and 1970—or they may not. If not, hydrogen engines may have to be introduced.

Sonic Motors

It may also be possible to make a change not in the energy source but the form in which it is put to use. Work done at Ohio State University by Professor C. C. Libby and others suggests that sonic (sound-powered) motors may provide competition for electrical motors in the not too distant future.

The principle can be seen in the crystals used in cheaper phonographs to convert into electricity the vibrational energy imparted to the needles. These are called piezoelectric crystals and, in the reverse process, will generate mechanical vibration forces when energized by alternating current. The result could be a motor with back-and-forth rather than rotational motion.

Claims by the researchers include: A higher efficiency (97 versus a typical motor's 80 percent); no moving (or rubbing) parts that are subject to wear or need for lubrication; simplicity of design, components, and assembly, leading to lower potential costs; reduction of 10 to 100 times in power requirements over those for the same processes when done by conventional motors; and the ability to do some jobs better, as in certain forms of riveting and metal forming.

Gravitation

One more possible variation on the energy theme is suggested by Dr. Joseph Weber of the University of Maryland. He points out that "There is about 100 times as much gravitational energy in the universe as nuclear energy; scientific studies of gravitation may yield the kinds of benefits to society which have come from other fields of scientific research."

13

Storage

In addition to the need for converting energy from one form to another, there may also be the need to store it for various periods. Indeed, one of the major reasons for conversion is that energy may be useful in one form but may store better in another.

Foods, fuels, and elevated bodies of water are the three methods of energy storage most used by man. (Every food is a fuel. If you would like a physical demonstration, try this: Place a cube of sugar in a shallow dish. Sprinkle the cube with pepper or nutmeg, then light one corner with a match. The sugar will burn with a hot flame! The sprinkled material is needed as a kind of kindling; otherwise the sugar simply melts.) Foods and fuels are both considered examples of chemical storage of energy.

It is probably fair to say that today the greatest need in the energy field is a good way to store electricity. (Hydroelectric pumped-storage plants, which we discuss later in the chapter, store water, not electricity.) Batteries, the only presently available method, weigh some 20 times more than the gasoline required to produce an equivalent amount of energy. (Depending on the system involved, the figure may vary from 5 to 200, but the lower figure involves some very costly devices.) And the cost of electricity from a standard flashlight cell runs about *100 times* that from a household socket. For some purposes, such as portable light, it is worth it. For standard electrical purposes—television, appliances, house lights, etc.—it is obviously out of the question.

Natural gas too is hard to store because it is so bulky. To obtain the energy equivalent of a gallon (.134 cubic feet) of gasoline would require that we store more than 100 cubic feet of natural gas. It can be compressed or liquefied, but this requires special equipment and adds to the cost. Presently it is stored in large underground or aboveground tanks, or in vast, porous rock formations. As with electricity, it has been found best to pipe it in the user's quarters, rather than store it there in tanks, such as oil tanks in homes and gas tanks in cars.

The major difference between natural gas and electricity is that natural gas is stored as such, even though not at the user's quarters. With electricity, it has been found best to create it at the time of use. But this is why public utilities are so concerned with peak demands, which can be at least double normal demands. Should they build generating capacity for peak demands, which may occur on only a few burning hot afternoons during the summer? Then large amounts of excess capacity are left idle much of the year, which is obviously uneconomical. It is interesting to note that before the widespread use of air conditioning, peak demands occurred during the winter, which left idle capacity during the summer. Public utilities began to promote air conditioning. The objective was "load balancing"; hopefully enough customers would install air conditioning to use up the excess capacity during the normally slower summer months. The idea was successful—indeed so successful that peak periods are now reached during the summer months. The tail is now wagging the dog.

While some utilities people would now undoubtedly like to "balance the load" by increasing winter use, others are becoming leery of this seesawing approach. Thus there is increasing use of gas turbines and pumped hydrostorage systems to meet peak demands.

Pumped Hydrostorage

Water, as we know, will only flow downhill. To be sure that there is always pressure in our water pipes, the stored water supply must be higher than the point of use. (And the greater the difference in height, the greater the pressure.) No matter how complicated and devious the route from storage to faucet, as long as the faucet is lower than the storage area, the water will flow.

Some communities are lucky enough to have this condition naturally. In other cases, or atop tall buildings, there will often be large water towers. Water is constantly being pumped up into the tanks and is stored there for use. Even though the water may be pumped up at a slow rate, the steadiness of the supply provides for peak demands even in the hottest weather.

So we see that water can be stored conveniently while electricity cannot. Combining the two provides us with what has come to be called pumped hydrostorage.

The idea is to use electricity generated during slow periods in a hydroelectric utility to pump water up into a higher reservoir.* Then, when peak demands must be met the

* The electricity may be generated on-site, or purchased from another site.

operator, in a sense, pulls the plug, allowing the water to run down and turn the turbines, thus generating electricity.

A pumped storage installation is about 65 percent efficient (as opposed to 80 percent in a standard plant; there is some loss in pumping up the water). This means that, over-all, 3 kilowatts of energy must be generated for every 2 that are eventually used. But the 3 kilowatts can be generated cheaply during off-hours, while the 2 can be used during peak hours and sold at a higher price. Thus hydroelectricity can be generated even where a hydroplant could not normally be built.

Although hydrostorage sounds like a fine idea, conservationists are up in arms about it. One large plant, planned by Consolidated Edison Company for Storm King Mountain 50 miles north of New York City to help supply the city's mounting demands, has been stalled for more than 7 years. Unfortunately, the very nature of such installations puts them into scenic areas; any region containing high areas overlooking a large body of water is probably a scenic area. And Storm King Mountain, a 1400-foot granite mountain, overlooking the mighty Hudson, certainly falls into this category. The Storm King plant has a planned capacity of 6 to 8 billion gallons of water and would generate 2 million kw of peak power. Whether the region would be ruined by cutting the necessary reservoir into the side of the mountain is presently being hotly debated, and is the major but not the only reason why objections have been raised. Among other problems: It will bring salt water further up the river, killing fish, and the electricity needed to pump the water creates air pollution elsewhere.

Another such project, a 1-million-kw installation at Northfield Mountain in Massachusetts, developed a leak in the reservoir and, as someone put it, filled up the middle of the mountain with water. One of the problems is that such higher areas often consist of rock that has been strongly weathered and fractured, and so does not make good reservoir material. Sealing methods have been developed but are far from foolproof.

An allied approach may help with some of these problems. In Sweden there are plans to build the world's first pumped-air storage power plant. Air is to be forced in a great underground cavern which is partially filled with water. The water is thereby pushed up a shaft to a surface reservoir. To generate power the cycle is reversed, compressing the air which is led to a gas turbine. Efficiency, however, is even lower than in pumped hydrostorage.

Flywheel

If a heavy flywheel is turned up to high speed, a fairly high amount of energy can be stored within it. This mechanical or kinetic energy can then be tapped as needed by a clutch arrangement, as is presently done in all cars when energy is taken from internal combustion engines. (Most cars already have a flywheel in the engine which helps to stabilize or smooth out the operation.) And flywheels are also used in some particle accelerators to store energy between pulses.

In principle the flywheel could be used as the sole power source. It would be spun up to high speed and could drive the vehicle for a given number of miles, after which it would have to be spun up again. This is very similar to filling up a tank with gasoline and knowing that you have, let's say, 250 miles of driving ahead before having to stop again for gas.

The problem with the flywheel is that putting that much energy into it is virtually impossible if it is to remain within reasonable size and weight. The Ground Vehicle Systems division of the Lockheed Missiles and Space Company has

HOW THE ENERGY PACK WORKS

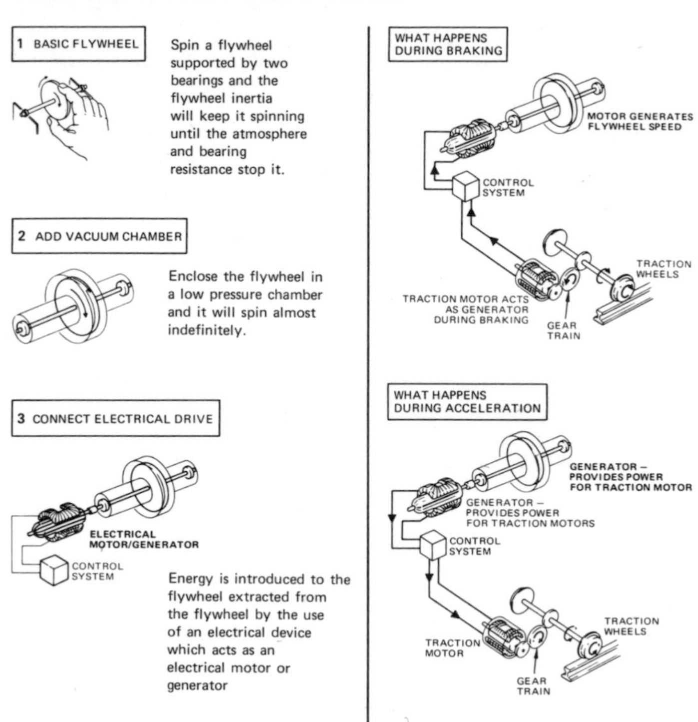

Proposed system using flywheel to conserve braking energy.

calculated that a flywheel to power a standard American car with all accessories for 200 miles would need a volume of 348 cubic feet (e.g., 8 feet in diameter and 1½ feet thick) and a weight of almost 3 tons! This is 50 percent more than the normal total weight of the car itself.

Strangely enough, such a device might be made to work in city buses, where the clean operation would be most useful, and where trips are normally short. Because the wheel could be spun up at the end of each trip, it would be smaller in relation to total size of the vehicle and might do the trick. Experimental buses powered with flywheels were introduced about 20 years ago in a few cities in Western Europe and the Congo. But they could only go about three-quarters of a mile before needing another "shot"; they were also complex devices to drive. New materials, controls, and knowledge has improved the picture since then.

Nor does this mean that the idea is definitely out for cars; the Lockheed people suggest that a hybrid system—a gasoline engine combined with a flywheel—might be a very good idea. The flywheel could be spun up in advance or during normal (nonaccelerating) periods. Another possibility is to use energy now dissipated during periods of braking! (A similar idea, shown in the drawing, is being looked into for the New York City subway system.)

Over-all advantages would be that horsepower requirements could be cut some 60 percent; the engine would not be required to operate beyond its most efficient speed. This would reduce fuel use and pollution, which are highest during periods of acceleration.

Another approach with the same objective in mind, but reminiscent of the pumped-air storage power plant, has been suggested by H. S. Dunn at the University of Rochester. In this case, peak power requirements are to be met by use of hydraulic accumulators that have stored up energy in advance. Such an accumulator consists of a strong rubber balloon blown up to some initial pressure, contained within a steel cylinder. During the storage cycle oil is forced into the cylinder, compressing the air in the balloon. When energy is required, the compressed air forces the oil out of the cylinder at high speed, which can then be used in some way, perhaps to turn a small turbine, and thereby supplement the base power supply.

The hydraulic accumulator might be combined with the

conventional internal combustion engine, but Dunn feels it would work even better with a gas turbine or Stirling cycle engine. The latter engine is one in which fuel is burned continuously (rather than intermittently). The heat produced is transferred to a confined gas that then actuates the pistons.

Heat Pump

In the chapter on solar energy we talked about heat storage using water, rocks, and even Glauber's salt. But there is another heat storage area we have not yet mentioned—the earth! Obviously the earth heats up each day as it is being pounded by the sun's rays. Is it possible that we can somehow make use of this stored energy? The answer is yes, and the method is called a heat pump.

A heat pump is a refrigeration device that is being used for heating rather than cooling. If you put your hand at the back or top of your refrigerator you will feel the heat that is being removed from the inside of the unit and spilled into the higher temperature of the room. In other words, a heat pump uses energy to transfer heat from an area of lower temperature (e.g. the ground or a well) to one with a higher temperature. Because the energy is going into *moving* the heat rather than creating it, a given amount of energy can put 2 to 3 times as much heat into the house as it could if used to produce the heat directly. Another advantage is that the same system can also be used for cooling by reversing the direction of flow of the pump.

It is this possibility that begins to make the heat pump an economical possibility, for it is much more complex and expensive than conventional heating systems. But with the large numbers of houses now being built with both electric

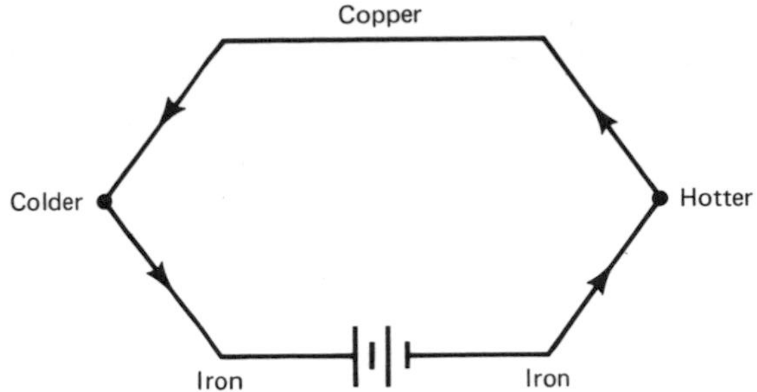

Thermoelectric heating and cooling using the Peltier effect.

heating and central air conditioning, the heat pump, particularly if made in large numbers, may begin to look more appealing in an economic sense. And if it becomes necessary to ration electricity, banning electric (resistive) heating, but permitting its use in conjunction with heat pumps, might be a good way to start.

It is also possible to construct a heat pump using the thermoelectric effect described in Chapter 3. For the inverse of the Seebeck effect also occurs. That is, if electricity is passed through a circuit made of 2 dissimilar wires, then, as shown, one junction becomes hotter and the other colder. This is called the Peltier effect; as with the device described earlier in this section, the Peltier effect too can be thought of as a heat pump: It is "pushing the heat uphill," the energy for which is coming from the electrical input, and may find use in residential and commercial buildings. An advantage of the Peltier effect is that changing from heating to cooling requires only a change in the direction of the current! Although we probably won't see this for a while, increases in efficiency are bringing the effect closer to the point where it might be used in household appliances, such as refrigerators.

Electrochemical Methods

The most common electrical storage system, the familiar flashlight cell or battery, celebrated its hundredth birthday in 1966. In all this time the chemical composition of G. L. Leclanche's original cell has changed but little, although it has been improved as a commercial product. The Leclanche battery and others of its type are called primary cells; after the chemicals are consumed the battery must be discarded. In recent years, however, new formulations have appeared which have increased the storage capacity of these devices by 5 or 10 times.

The most important characteristic of so-called secondary or storage batteries is that they can be recharged by passing electricity back through them. The most common type is the lead-acid battery still used in automobiles. In addition, smaller recharageable types have been developed for use in portable appliances such as electric carving knives, shavers, and the like.

The major requirement today, however, is for rechargeable, high-capacity devices. The most important application would be for powering automobiles. Although some advances have been made, a practical, economical device remains to be developed.

The lead-acid type is reliable and relatively inexpensive, but cannot hold much charge. And therein lies the major problem. A number of experimental cars powered by lead-acid batteries have been put together. In general, they are limited to speeds of about 35 miles an hour and ranges of about 30 to 50 miles, while the battery weight is often equal to the weight of the rest of the car.

Electric systems are ideal for the stop-and-go traffic of cities, however, and especially for multistop delivery service. For when a stop is made, the engine is not drawing any power at all. An electric truck made for this purpose, the Battronic,

Designer-builder Bob McKee shows how battery pack of his 8-horsepower "Sundancer" can be changed rapidly.

has a maximum loaded weight of 9500 pounds, of which 2000 to 3000 are batteries (depending on range and speed desired). Small commuter cars are another likely possibility.

Storage capacity is most often measured in watt-hours per pound, or energy density. Typical (and approximate) figures for various systems are given below, where P = primary battery, S = secondary cell.

Energy Source	*Energy Density (watt-hours per pound)*
Flywheel	25
Leclanche (P)	30
Lead-acid (S)	10

Chart continues on following page

Energy Source	*Energy Density (watt-hours per pound)*
Zinc-air (S)	80-100
Lithium-sulfur (S)	100
Molten sulfur (S)	150-200
Thermoelectric generators	200
Internal combustion engine	600
Hydrazine fuel cell*	300-500
Metal hydride fuel cell	1000
Hydrocarbon fuel cell	2000

* Discussed in the next section.

It can be seen that the less well known types have already considerably exceeded the lead-acid battery in energy density, and some are already in use for military and space applications. But in general they all have some penalty associated with them in terms of consumer use, such as high cost or danger. Batteries using molten sulfur or other such materials, for instance, must operate at very high temperatures, perhaps 900° F. Aside from the potential danger, there is the problem of maintaining this high temperature during periods of nonoperation, or bringing the unit back up to operating temperature when it is needed.

The Atomic Energy Commission, in its first major foray outside its usual area (based on recent Congressional authority to expand its scope), is looking into this question, and is doing work on the lithium-sulfur combination which, hopefully, will be able to bring auto speed up to 50 or 60 miles an hour and the range up to 150 to 200 miles.

If satisfactory battery systems can be developed, other areas of use would be open to them. A most important one would be that of storing electricity to meet peak demands for electric power. Thus the need for the controversial pumped storage plants would be lessened or eliminated.

Other advantages are that the storage systems could be placed closer to the source of demand (assuming that there would not be one in every home), and that they could be installed in weeks or months rather than years (or even decades, considering the technological and public pressure problems that power plants seem to be facing today).

Fuel Cells

A somewhat similar device that has been making news in the last decade or so is the fuel cell, which someone once described as a continuous-feed battery. That is to say, it is a device in which the chemicals involved (fuels and oxygen) are steadily added and the by-products constantly removed. A simplified hydrogen/oxygen fuel cell is shown.

In essence a fuel cell evolves water and electrical energy,

Schematic diagram of hydrogen/oxygen fuel cell.

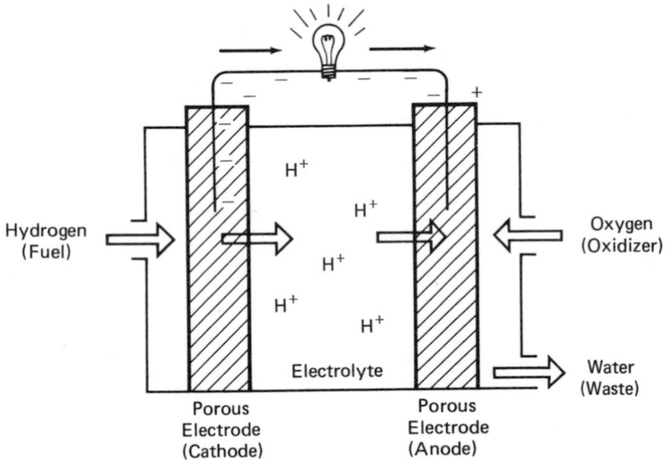

reversing the process of electrolysis. Batteries, on the other hand, store electricity that has been generated elsewhere.

A main advantage of the fuel cell is that, like the direct conversion methods we mentioned earlier, it bypasses the turbine-generator cycle, and so promises higher efficiencies. Because it operates at a constant temperature, its efficiency is not limited by the difference in working temperatures, and so the efficiency can be very high, typically 50 percent, and theoretically approaching 100 percent.

Although the fuel cell is actually a conversion rather than storage device, we include it here because of its general similarity to batteries. Like a battery, it uses direct chemical reactions to make electricity. Unlike a battery, it doesn't wear out and doesn't need recharging.

Some of the newer batteries are able to provide high peak power and heavy service, though they have limited energy density. Fuel cells offer the reverse combination. This means that fuel cells, when used for vehicles, will probably have to be hybrid systems—combined with batteries—to provide for peak power needs.

Although the fuel cell was invented back in the 1830's, technical problems held back development until our space program provided a strong stimulus to further development. The use of fuel cells in the Gemini and Apollo programs has been well documented. In recent years interest has otherwise waned, however, partly because of the high cost of the component parts. Very high-cost catalysts,* for example, have been needed in the electrodes to make the process work. In addition, some of these, e.g., platinum, are in very short supply.

Two main development programs remain in operation. One is aimed at reducing the dependence on, or the cost of the

* Substances that speed up chemical reactions without themselves entering into the reactions.

catalysts and is being supported by the Army for off-road vehicles. Clearly the silent, reliable operation of such devices (the vehicles are actually powered by electric motors, which have only one moving part) would be very desirable for the military. This may one day provide the answer to the increasing problems of automobile noise and pollution.

The other major program, being supported by private industry, aims at the direct conversion of natural gas to electricity right in the home. Proponents point to the fact that natural gas can be stored, and that transmission costs are lower than for electricity. The photo shows a prototype

Experimental natural gas fuel cell power plant in a home. At left, unit houses fuel cell power section and a "reformer." At right, unit houses inverter, which converts DC to AC.

fuel cell power supply actually installed in a home. It is quiet and practically pollution-free.

On the other hand, it has also been pointed out that in a given area, the various electric appliances in different homes are being turned on at different times, and some are not turned on at all. This means that a utility can average out the load—that is, it can satisfactorily supply the area with an average of perhaps 500 watts per home, whereas with individual units, fuel cells with ratings of several thousand watts are required.

Further, as the photo shows, a separate unit called a reformer is needed to convert the natural gas to a form that is useful in the fuel cell. This raises the cost and cuts the efficiency down to 35 to 45 percent, or in the range of conventional power supplies.

Development may still bring lowered catalyst costs and use of hydrocarbon fuels without reforming. And of course, air pollution requirements may make such a move a requirement, while a change to a hydrogen system would bring the entire approach much closer to practicality. Perhaps substations of several megawatts, each serving small regions, would be the most sensible approach.

It should be pointed out that there is not one, or even one type of, fuel cell. A wide range of fuels, operating temperatures, pressures, and designs create a vast range of possibilities.

Should any of these approaches work, we may then see the use of fuel cells to power practically everything—boats, cars, locomotives, construction equipment, and other vehicles, as well as providing electricity for residential, industrial, and commercial use.

Fuel cells also lend themselves to interesting combinations with other technologies. One that we have already mentioned is the hydrogen economy. Another is the use of synthetic gas manufactured from coal, to supplement or replace natural gas.

In space every ounce counts. Thus the by-product of the fuel cell, namely water, has been processed and reused. But fuel cells may also provide the answer to the disposal of human waste, by utilizing the chemical energy still contained within it. (Recall, for example, that cow dung can be burned.) Another intriguing possibility is the use of small, implanted biological fuel cells for use as heart pacemakers or even power for artificial limbs, which would run off materials found within the body.

Obviously it would not make sense to try to use the electricity to convert the water back into hydrogen and oxygen. The laws of thermodynamics tell us we would lose out in the process. But suppose there was some waste energy we could put to work? J. O'M. Bockris of the University of Pennsylvania suggests, for example, the use of radioactive nuclear wastes to perform this regenerative process. If the fuel cell were made a closed system—simply converting the hydrogen/oxygen back and forth from water—little or no radioactivity would escape.

And finally, M. D. Rubin, a consulting scientist, suggests that "an attractive though sophisticated combination for producing heat efficiently would be the combination of a fuel cell with a heat pump, with probably the greatest coefficient of performance available."

Nuclear Storage

The various kinds of storage we have mentioned, plus one other we have not yet discussed, can be arranged as shown below. To date, our greatest capability in terms of reuse lies in the topmost level, the mechanical. In the chemical area we have devices that store energy but they are either expensive or inefficient, though present-day research is certainly improving the picture.

Levels of Energy Storage

Type	*Example*	*Energy Density (Watt-hrs/lb)*
External (mechanical)	wound clocksprings	.05
	twisted rubber bands	7
	flywheel	25
Molecular (chemical)	storage battery (lead-acid)	10
	frozen water	42
	condensed steam	282
Atomic (combustion)	burning firewood	1850
	burning alcohol	3500
	burning crude oil	4700
	burning rubber bands	5700
	burning hydrogen gas	15,100
Nuclear	fission of the ^{235}U in 1 pound of natural uranium	73,550,000
	fission of 1 pound of uranium-235	10,000,000,000
	fusion of hydrogen to 1 pound of helium	75,100,000,000
	complete annihilation of any matter	11,320,000,000,000

In the area of combustion, energy density soars. Note that burning rubber bands gives 800 times as much energy as twisting them. But this is not really a fair comparison because once used, the fuel is gone. As we noted earlier, we may one day have to go to fuels that can be produced today, such as wood or alcohol.

The levels of energy obtained from matter depend upon how deeply we penetrate them. The mechanical use is an external one and depends on the device itself—an aggregate of molecules. In the next level we are utilizing the bonds between atoms and molecules. The third level, the atomic, depends more upon the electrostatic attraction between nucleus and electron. And then, lastly, there is the nuclear level.

In terms of primary energy supplies, the nuclear level is exemplified by fission and fusion. But in this chapter we are talking of storage. Radioactive materials can be thought of

In the nuclear battery shown, a central rod is coated with an electron-emitting radioisotope, for instance, strontium-90. The electrons emitted by the material speed across the gap to the collecting cylinder surrounding it, thus generating the required potential (or voltage) difference.

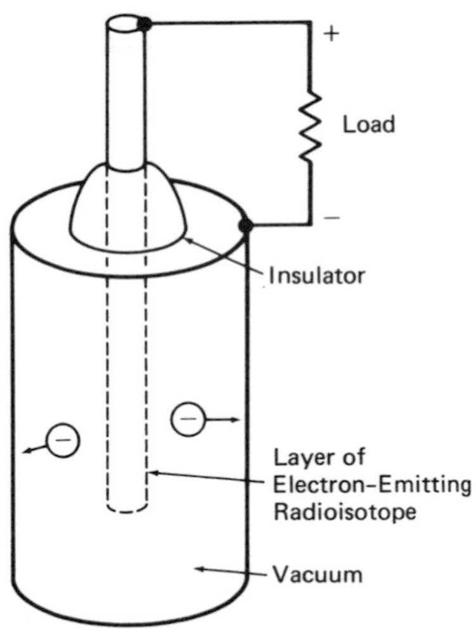

as containing stored energy, which is easily gotten at; and we can even produce them at will. Thus we can begin to talk of nuclear batteries; indeed, some experimental models have been built. They utilize the electrons emitted by radioactive materials to build up a charge. Thus far devices of this type can develop high voltages but only very little current. Potentially the energy density can be high, and it is likely that scientists will someday be able to exploit this area.

A slightly different approach to nuclear storage involves use of a space station, which would store solar energy in the form of high-energy particles. When such particles strike suitable targets they produce uranium-233 and plutonium. These fissionable fuels could then be transported back to earth and used in generating stations. If done on a large enough scale, the need for breeders might be eliminated or at least reduced.

14

Transportation of Energy

TRANSPORTATION IN THE United States uses one-fifth of our Gross National Product. And of the vast bulk of goods we transport, fully half is fuel! Then, in addition, there is a large capital investment in movement of electrical energy.

In size alone it can be seen that the transportation of energy makes up an important part of our economy. But advances in technique have also had important effects on other aspects of our lives. Earlier improvements in electrical transmission meant that generating plants did not have to be quite so close to the population load, leading to sitting of power plants in more remote areas, as well as the building of a large number of dams and hydroelectric plants. Newer improvements in energy transmission may make it possible to build more mine-mouth plants, to eliminate unsightly electric lines and utility poles, to put nuclear power plants in orbit (if that should prove desirable), to make (batteryless) electric cars a reality, and, well, feel free to make suggestions. High or low transportation costs could spell the difference between shortage and sufficiency, at least for a while.

In general, energy can be transported in one of four ways. It can be shipped in bulk, as a trainload of coal, a tankload of gasoline, a cord of wood, or a pound of uranium. Or it can be piped; perhaps surprisingly, all three fossil fuel types—coal, oil, and natural gas—can be shipped by pipeline. (The coal is ground down and mixed with a fluid.) A third way to transport energy is via wires in the form of electricity. And the last way, still experimental, is to ship it via electromagnetic radiation—microwaves and laser light are just two possibilities. In this chapter we shall look at these methods, but from the point of view of the individual kinds of fuel and energy. We start with that Cinderella of the fuel world, coal.

Coal

We have seen that coal, our only abundant fuel, is making a kind of comeback. And if gasification becomes a commercial success, transportation of coal (e.g., to gasification plants, which may not necessarily be at the mine mouth) may grow rapidly in importance.

In earlier days the cost of transporting the coal was often higher than its value at the mine. Transportation still accounts for perhaps one-third to one-half of a utility's fuel cost. Among the developments that we have seen or will see are: The unit train, which consists exclusively of coal cars that operate on a fixed route; long-distance conveyors; and pipeline delivery. In the latter two methods there is no need for empty "returns." A more advanced concept of the unit train is the integral train, consisting of extra large cars with power units interspersed between them to eliminate the need to turn around or decouple the engine, and specially designed for rapid loading and unloading.

Oil

For movement of oil overland, pipelines are generally the cheapest method. And the larger the pipeline, the lower the cost. Pipelines up to 42 inches in diameter have been built. They have been put underwater, in polar regions, and can go up and down mountains with not too much more trouble. Petroleum people claim that new methods of insulation have made it possible to ship oil at 180° F. through a pipeline without even melting snow that has fallen on it.

Environmentalists are not convinced of this and fear that a hot pipeline (hot oil is less viscous and travels more easily through the line) may disturb the permanently frozen ground of Alaska if the controversial trans-Alaska pipeline is built. Another objection is that the line would travel through several earthquake zones. Oil spills are bad enough in temperate and hot areas where the effects may last for months or years. But in the Arctic, where petroleum-hungry microorganisms can only operate for a very short period each year, the effects might last for centuries. Among other problems is the possibility that the line might interfere with migration routes of caribou and other Arctic creatures.

Proponents of the line concede problems, but maintain that if it is designed and built with care, all these problems can be taken care of and that, more important, the large oil and gas discoveries in Alaska could help lessen our growing dependence on foreign oil. The proposed line would be almost 800 miles long, 48 inches in diameter, and capable of eventually transporting 2 million barrels of oil a day from the oil fields of the North Slope to the port of Valdez in southern Alaska. From there it would be transferred to tankers to complete the journey. The alternate all-tanker route requires travel through the Arctic Ocean, which is frozen a large part of the year and dangerous at all times. Unfortunately, the tanker part of the route is another danger area in terms of pollution (particularly oil spills), with Prince William Sound, on which Valdez is located, containing extremely strong currents and winds.

The very large tankers that would be used—250,000 tonners are now in use and 500,000 to 1,000,000 tonners are planned—are not exactly agile. A typical large tanker, say a 120,000 tonner, has a turning diameter of half a mile and an emergency stopping distance of 2 miles.

The Japanese *Nisseki Maru*, a 373,000-ton supertanker, is almost a quarter of a mile long and carries 120 million gallons of oil; its anchors weigh almost 25 tons apiece and the paint that covers the giant ship weighs 300 tons.

The supertankers have begun to assume a big role in international energy transportation. This is mainly because the largest deposits of oil are not in the countries that need them most, and the cheapest way to ship oil is by tanker.

Further, the economics of overseas transportation calls for large tankers, but past problems with oil spills have made the American public wary of these huge ships. This is one reason why we have few of the deep-water ports needed by them. As an indication of what we face, the largest tanker the Delaware River can accommodate is 80,000 tons, and the San Francisco oil port only 70,000 tons. Yet 85 percent of the tankers under construction or on order in 1971 had capacities of over 200,000 tons. Only one area exists along the entire East Coast capable of handling even this size, on the northeast coast of Maine. Clearly some changes will have to be made.

Alternatives that we might see later on would be some form of icebreaker tanker to move across polar routes that are thousands of miles shorter than normal trade routes, and perhaps even submarine tankers which would eliminate both ice and weather problems.

Another possibility is an air bag that can be filled with a petroleum product and towed, then folded up for the return trip.

Natural Gas

Natural gas can be, and is, shipped by pipeline. Much of the gas used in the Northeast is piped from the gas fields along the Gulf Coast. One of the main reasons that the use of natural gas has grown so extensively is the large distribution network that has grown up.

Now, however, use has begun to outrun supply, which means we must begin to import. But no one has figured out a way to build transoceanic pipelines, which means tankers must be used. Unfortunately, while shipping gas by pipeline is relatively cheap, its great bulk makes shipment by tanker expensive. Alternatives are compression and liquefaction.

Gas can be compressed into heavy-walled steel tanks and for some kinds of use this works very well. Containers of compressed gas are commonly held to less than 250 pounds per square inch, which represents a reduction in volume of about 15 to 1.

But the boiling point of natural gas at atmospheric pressure is −259° F. If the gas can be brought down to that temperature it liquefies, in which state it occupies 1/632 the corresponding gas volume. Thus, for long distance, high-volume shipment liquefaction (to liquefied natural gas, or LNG) has begun to look increasingly interesting. What is involved is a liquefaction plant at the sending country, a tanker with well-insulated tanks that can maintain the cold temperature, and regasification and transmission facilities at the receiving end.

Shipment of LNG overland by truck and tank car is com-

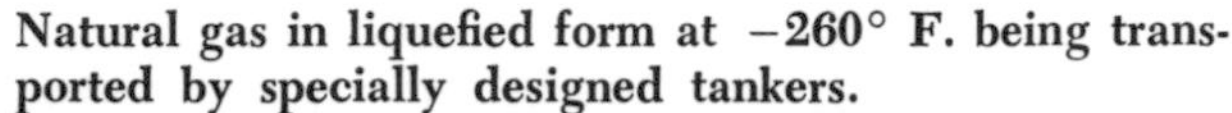

Natural gas in liquefied form at −260° F. being transported by specially designed tankers.

mon; but shipment by pipeline would be very expensive at this time, for it would require refrigeration of long lines of pipe, as opposed to the smaller surfaces involved in trucks and tank cars. Refrigeration of the lines does offer large advantages, however, such as needing only $\frac{1}{10}$ the horsepower requirement for pumping the material through the line.

Another alternative we may see in the future is the production from natural gas of a product called methanol, a liquid at room temperature and pressure, which nevertheless burns as cleanly as natural gas. The advantage is that conventional tankers and pipelines can be used for shipment; unfortunately as with LNG, the additional work required adds considerably to the cost of the product. The Federal Power Commission has just authorized the first long-term import of large quantities of LNG. It is expected that a fleet of 100 large refrigerated tankers may be plying the seas by 1990, each carrying the equivalent of some 2.7 billion cubic feet of natural gas.

Transmission of Electricity

Even though electricity is the most expensive form of energy transport, particularly for distances over 500 miles, the rapid growth of the electrical world has carried with it an equally rapid growth of electrical transmission capability and called forth some new developments as well. In 1970 some 4 million acres of land were devoted to transmission facilities (including rights of way for electrical lines); by 1990 this right rise to 7 million acres, or 17,000 square miles —the combined areas of Vermont and Massachusetts. This could prove to be more of a problem than siting of power plants.

Several advances, present and future, may hold this down,

however. One line can be made to do the work of two if the voltage of the line is upped. Maximum overhead line voltages have increased from 345,000 volts in the 1950s to 765,000 volts (765 kv) today. Although such lines may be double in cost, they can transmit 3 to 4 times the power, and over considerably longer distances.* Such extra-high-voltage (EHV) lines have made feasible the movement of large blocks of power between neighboring power systems. Thus when one system has trouble due to a breakdown or other overload it can borrow capacity.

Lines in the 1000- to 1500-kv range are considered quite possible, but whether economical or even feasible is another question. There would be problems of clearance for instance. (J. G. Cline, chairman of the New York State Atomic and Space Development Authority, tells us that the clearance required for high-voltage lines was determined years ago as the amount of space needed by a farmer atop a hay wagon holding a pitchfork on his shoulder.) Other environmental problems connected with EHV lines are: more massive and taller towers, thicker (more visible) wires, wider rights-of-way, and greater corona (electrical discharge).

It is likely that, especially for long-distance movement of large blocks of power, DC rather than the common AC lines will be used. The Soviet Union, even more spread out than the United States, has pioneered in this field. Direct current lines have less losses over long distances, are less expensive, and are easier to bring up to high voltages. On the other hand, they require conversion facilities at both sending and receiving areas (because both generation and use are in AC).

The shipment of electricity, it is worth noting should be the most economical method of shipping energy, mainly be-

* One can figure, approximately, a need of 1000 volts of "electrical pressure" for each mile of transmission distance.

cause of the low mass of the unit being shipped, namely, the electron. At present, however, it is still several times as expensive as pipelines. Shipment of electricity over 500- to 765-kv lines is said to be almost competitive with equivalent shipment of coal. It also makes possible the shipment of large blocks of power, say, north in winter and south in summer. This could cut reserve requirements considerably. In any case, the larger power stations being contemplated—perhaps on the order of 5 million kw—will require use of these extremely high voltages, at least for a few major trunk lines.

In those cases where power plants are (or are to be) located in or near cities, another problem arises. The public has insisted that power plants not be sited within populated areas. The public has also stated its displeasure with disfiguration of its countrside by mile upon mile of overhead transmission line, along with wide swathes cut through forest and field.

Even if perfectly clean and safe power plants are developed, so that they could be safely located within city limits, an extensive network of wires is needed to distribute the electricity to the various users. How are these conflicting considerations to be meshed?

Clearly the best answer is to put the transmission and distribution systems underground. Unfortunately, this is far more expensive than stringing wires up along poles. Presently only about 2400 miles of transmission lines, out of a total of some 300,000 are buried.

The Electric Utility Task Force on Environment observed a few years ago that the replacement of all overhead distribution lines would require an investment on the order of $150 billion. Nevertheless, the power industry has generally supported the idea of placing all new residential subdivision distribution lines underground.

Underground transmission of bulk power presents problems of a different order entirely. At present, no satisfactory method exists for shipping high-voltage power underground for more than about 20 miles. In spite of all precautions that can be taken, a 345-kv underground cable generally has less than half the capacity of a similar overhead line because of a buildup of heat. Commonwealth Edison Company of Chicago has recently increased the capacity of 138-kv cable 50 percent by adding an oil circulation and refrigeration system. A further improvement may come with experiments now in progress to cool the lines with compressed gases such as nitrogen or sulfur hexafluoride.

The heat, of course, is caused by the fact that even though the metal used (usually but not always copper) is a good conductor, it still offers some resistance to the current. This resistance shows up in heat. In the approaches to the problem mentioned so far, the objective has been to remove

Proposed 3-phase cryogenic cable system.

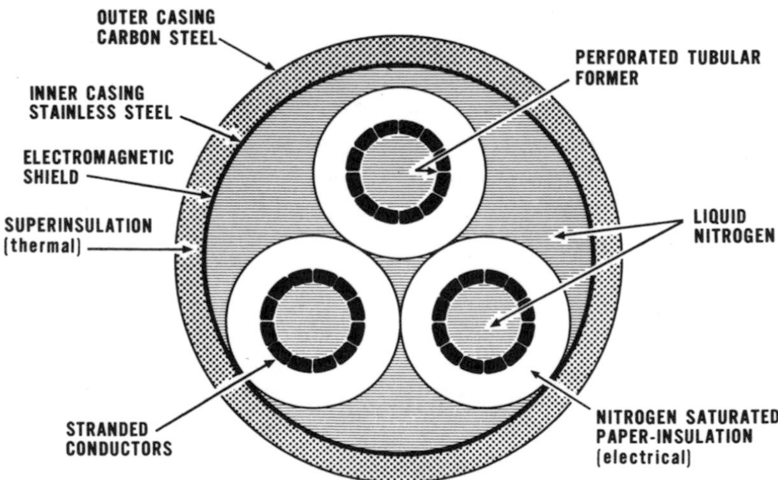

the heat. A more direct approach is to cut the resistance of the line. This can be done by cooling it to the point where it becomes superconducting. Although this requires extreme cooling—down to 4.2° F. above absolute zero (or −455.2° F.) for a superconducting material called niobium, a 230-kv system would have 8 times the carrying capacity of a typical 345-kv line, and the cost, strangely enough, is estimated to be only about 60 percent of one of today's conventional underground lines. A superconducting DC line is another possibility. It has been estimated that three 17-inch pipes could carry the entire power needs for all of New York City!

But once a trench is to be dug, or a tunnel drilled, the difference between putting in a 17″ or a 30″ pipe, or 2 or 3 pipes, is not all that great. It may be possible to make the entire process even more attractive, once it is fully developed, by using the pipes for other purposes as well, such as shipment of liquefied gas. This could well be called an energy pipeline. Electrical transmission would take place through a cable kept at superconducting temperatures by, let's say, liquefied hydrogen and/or methane. The low-resistance cables could probably also be used for communications, particularly high-volume data communications, another field that is growing rapidly.

If high-capacity superconducting lines become a reality, we may see a complete restructuring of our power-generating system. One possibility is location of all power plants off the coasts of the country, with power grids carrying the necessary power to the various parts of the country.

Electromagnetic Radiation

A final possibility in energy transmission is the use of electromagnetic waves. Energy in the form of radio and microwaves has been sent through pipes (called wave guides), the atmosphere and space for many years. Experiments have shown that microwave energy can be used to actually transmit power through the atmosphere, as in the case of a model helicopter whose power was derived in this way. Laser beams are another possibility. At the moment none of these methods is actually possible (except for the wave guide), both because of loss through spreading (even though laser beams spread very little) and through blockage by the atmosphere. There is also the danger of potentially injurious effects on living things. But I have learned never to say that something is impossible. We may yet have nuclear plants in orbit around the earth, transmitting their power to earth via microwave beams.

15

Conservation

AMERICANS HAVE BEEN brought up on a diet of cheap and plentiful energy, and have as a result developed wasteful habits of consumption. Lights are left on in empty rooms and even overnight. Our houses are overheated, often to the point where windows have to be opened. Big, heavy cars that are veritable gas guzzlers are the rule rather than the exception. And we have frost-free refrigerators with *heaters* in them to melt the frost.

Washing machines, refrigerators, and other major and minor appliances are often simply thrown away after a few years when a new model is introduced. The discarded items then become wastes that must be carted away and somehow gotten rid of. All the energy that went into making these items is then wasted, as are the materials of which they were made.

Perhaps we should begin to think in terms of borrowing materials from a common "pot," to which they would be somehow returned. There are several possible approaches.

Recycling

First and most obvious is that of recycling materials. This is certainly not a new idea. It has been done for years in many industries. As a matter of fact, as the economy "grew up," we saw the demise of a certain recycling programs that had been taken for granted for years. The New York City Bell Telephone Company used to collect the old directories when delivering the new one. No more. Reason? They had trouble finding buyers for the paper. Milk and soda bottles used to be collected and a few cents deposit paid for each one. This too has disappeared in favor of no-deposit no-return bottles (and of course a much expanded use of cans). Reason? No one wanted to be bothered with the extra work of sorting and handling. In other words, we have become spoiled. We almost automatically opt for energy-intensive rather than labor-intensive or even materials-intensive processes.

It seems fairly clear that this will have to change. It requires only 700 kw-hrs of electricity to produce a ton of steel from reclaimed steel, as opposed to 2700 kw-hrs to produce it from ore. For aluminum the figure is 5 times as much. And even if glass is collected and reprocessed instead of thrown away, it takes 4 times as much energy to melt down no-return bottles and make new ones as it does to collect, clean, and refill them.

R. S. Berry, professor of chemistry at the University of Chicago, is concerned about "built-in obsolescence," and says cars today are virtually designed to last only a few years. Professor Berry feels that making cars last longer could result in energy savings of 50 to 60 percent, as opposed to savings of perhaps 20 percent from recycling. He adds, however, that both approaches are valuable and should be taken, and puts them both under the heading of "thermodynamic thrift."

At the moment, economics and convenience still dictate to a large extent our throwaway society. Even perfectly sound attractive buildings are often torn down because the owner feels he can make more money from a different kind or size.

In my town we have a fairly well established system for collection of glass, paper, and aluminum. But, except for one driver, it is all done by volunteer labor, with the organizations supplying the manpower collecting the relatively small

Refuse production per person in the U.S.

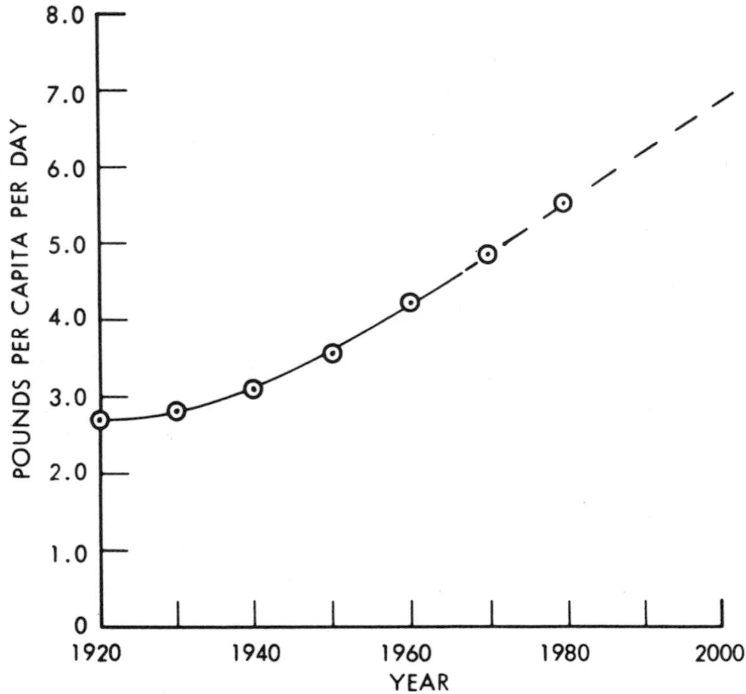

monetary returns. The town donates the use of one of its large trucks and a church the use of its parking lot. Could the whole thing be made to pay off if all proper expenses had to be paid? The town council is looking into the question now to see if the operation can be put on a permanent working basis.

Waste Is a Valuable Commodity

Thermodynamic thrift would help solve another major problem as well. That problem is the mountains of trash that are accumulating everywhere across the country.

Recycling not only saves materials and energy, but can be made to pay off. A proposed system by the Aluminum Association.

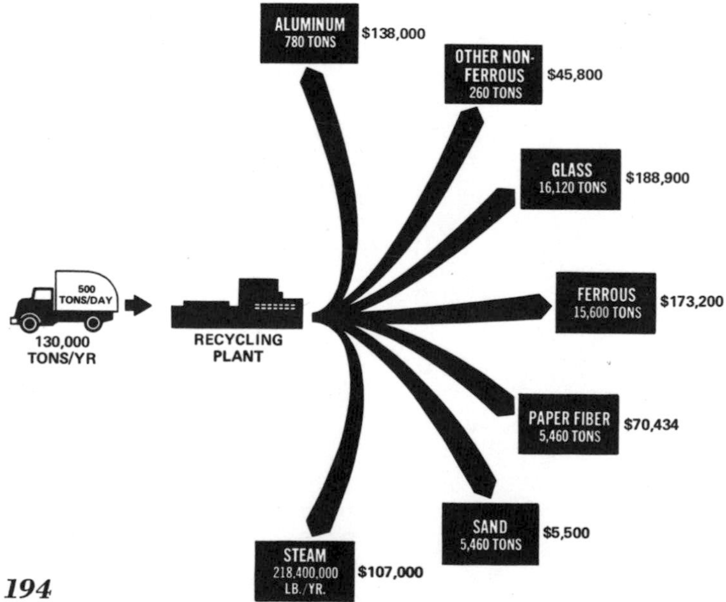

Conservation

We saw earlier that it is possible to reclaim certain materials from the smoke of fossil fuel power plants. It turns out that while smoke contains valuable elements, trash contains energy, indeed, quite a bit. Surprisingly, the fuel value of municipal solid waste is on the order of half that contained in coal!

The most direct way to get this energy out of the trash is to burn it in a controlled fashion. The idea is simple in concept, but more difficult in practice. Nevertheless, it is being done, and valuable quantities of electricity and steam are being obtained thereby. This is fairly widespread in Europe, far less so here.

In one pilot program being run in St. Louis some 300 tons a day of trash (about 30 percent of the total collected) are ground up mixed with coal, and burned in a power plant boiler. Prior to breaking down the refuse to 1½″ size, ferrous metals (iron and steel) are removed magnetically. Otherwise all materials are ground up together. The materials that do not burn are simply removed along with the ash.

A major bonus: Refuse, which is to be mixed with the coal in the proportion of 10 percent to 90 percent coal, is low in sulfur!

At some later date is is possible that other removal schemes will make it possible to economically reclaim other materials. S. David Freeman of the President's Office of Science and Technology, maintains that "our solid wastes by and large constitute a better grade ore than many of the ones we mine. If we reclaimed the iron, lead, aluminum, and other metals in our trash, using only a small fraction of the energy required in the production of the metals from the raw ores, we could help satisfy future demands. Thus, we could conserve our energy resources as well as our metals through recycling." How? Mr. Freeman (and others) suggest that we adjust the price of energy to include the damage its use does to our environment. At higher prices, users will quickly find ways to recycle, he points out.

Among the more imaginative, and daring, suggestions along this line is one put forth by two Atomic Energy scientists, William C. Gough and Bernard J. Eastlund, who wondered how the great energies promised by fusion power could be put to use. Their answer is the "fusion torch," which we might call the ultimate in incineration. Because the temperatures of fusion are higher than the vaporization point of all known elements, trash and garbage fed into such a device would automatically be turned to gas. At the very most there would be 92 different natural elements that would have to be separated out; techniques already are available to do this, at least on a laboratory scale. The result would be an automatic sorting out of all the basic elements, rather than the complications of sorting out a complex mixture of substances and compounds, such as plastics, metal alloys, etc. (Another possibility is to get rid of nuclear wastes, not by burying them or sending them to the sun, but rather by nuclear transformations.) The fusion torch has been called "abso-

Diagram of a fusion torch.

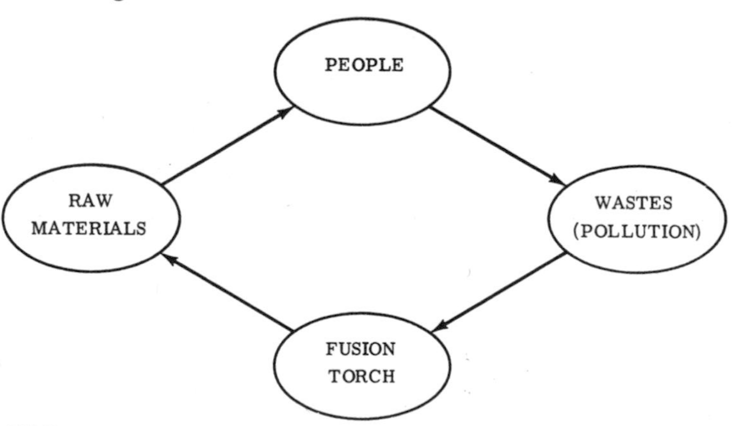

Average Composition of Municipal Refuse (% by Weight)

RUBBISH (64 PER CENT)	
PAPER, ALL KINDS	42
WOOD AND BARK	2.4
GRASS	4.0
BRUSH	1.5
CUTTINGS, GREEN	1.5
LEAVES, DRY	5.0
LEATHER GOODS	0.3
RUBBER	0.6
PLASTICS	0.7
OILS, PAINT	0.8
LINOLEUM	0.1
RAGS	0.6
STREET REFUSE	3.0
DIRT, HOUSEHOLD	1.0
UNCLASSIFIED	0.5
FOOD WASTES (12 PER CENT)	
GARBAGE	10.0
FATS	2.0
NON COMBUSTIBLES (24 PER CENT)	
METALS	8.0
GLASS AND CERAMICS	6.0
ASHES	10.0
	100.0

lute nonsense." Perhaps it would be better to call it a method for the distant future.

But there are other ways of utilizing trash that are much closer to realization. The fact that trash can be burned tells us that it contains hydrocarbons. A number of ways have been found to either extract or convert them into usable form. Heating trash without the presence of oxygen, for example, has been found to produce a usable grade of either oil or gas, or both. Similar results have been produced by fermentation. We have already mentioned the use of fermentation to produce alcohol, a burnable substance. In similar fashion, one-celled organisms have been found that dine happily on such tidbits as old tires (which contain carbon). The Firestone Tire and Rubber Company is actually building a plant that will produce oil and gas from old tires.

The largest contributor to our waste problem is agriculture. American livestock produces over 2 billion tons of manure each year, which is becoming an odor, health, and pollution problem. While the manure makes a good fertilizer, there are serious sanitary problems involved in the handling. It has long been known that vast amounts of methane, sometimes called "swamp gas" or bio-gas, are generated each year by the natural decomposition of organic material. Methane is the principal constituent of the natural gas that is pumped out of the ground by gas companies, but because of our wasteful habits, the potential of bio-gas has been ignored up to now. Not so in India, where small-scale plants have been put into operation to produce from 100 to 9000 cubic feet a day of bio-gas from cow manure, of which there is an ample supply. The result is that many village women, who formerly had to cook with dried manure, can now cook with gas.

Professor H. L. Bohn of the University of Arizona maintains that in the United States conversion of animal wastes to methane "could supply a clean fuel that could almost double the country's supply of natural gas." * "How about a home," suggests Sheldon Novick, editor of *Environment*, "in which garbage and human wastes are converted into natural gas which is then used for domestic needs (there

* *Environment*, December 1971, p. 5.

wouldn't be enough, but we can allow some outside fuel supply)?" *

A team of researchers at the Pittsburgh Energy Research Center have figured out a scheme for turning cow manure into oil. They believe that an oil yield of 50 percent of the dry weight of the manure could be realized. It has even been suggested that on long space trips (where any kind of waste is intolerable), human wastes could be turned into rocket fuel!

Another large contributor to our waste problem is cellulose, a by-product of our wood and paper industries. Urban wastes also contain great amounts. These wastes too are potential resources for microbial conversion to fuels. A number of such projects are in the works across the country.

The process is basically simple. The organisms that do the work are already present in the waste materials. It is only necessary to provide a benign environment for them—meaning exclusion of oxygen, a watery mixture of the wastes, circulation of the mixture, and warmth (a temperature of about 95° F.). Along with the gaseous product, a practically sterile solid material is produced that can be used as a soil conditioner or fertilizer.

A final thought along these lines: Among the substances used by plants are nitric acid and other nitrogen compounds. Surely it would be possible to find a way to take what are now pollutants, oxides of nitrogen, and convert them into a useful substance. The high temperatures involved in magnetohydrodynamic production of electricity may make such a step relatively easy to accomplish. Nevertheless, the extraction costs would probably be high.

* *Ibid.*, p. 3.

Creature Comforts

We often see huge walls of glass in office or apartment towers admitting great quantities of heat while giant air conditioners labor mightily to remove it and dump it outside, thus making the surrrounding air even hotter; and all the while the electric utility's generators struggle just as hard to keep up with the load.

Not only individual buildings but entire cities can be properly oriented with respect to the sun, and designed and constructed to take advantage of prevailing breezes, shade trees, and so on. With such ideas in mind, we might find fewer poorly ventilated, low-ceilinged houses being built that require more or less continuous air conditioning in the warm months.

As with the office buildings mentioned earlier, it is simply easier and more convenient to set the air-conditioning thermostat and then forget about the weather. But the waste of energy is enormous and, with the rapid growth of air conditioning, is getting worse.

Similarly, we have forgotten some basic principles in keeping our buildings warm. Just as great walls of glass are not a good idea in hot, sunny weather, so too are they a bad investment in terms of cold (except during sunny winter days, and then only if the windows are properly oriented). Even the thick double-paned, so-called insulating glass comes nowhere near the insulating qualities of masonry or insulated wall construction.

Federal Power Commission chairman John N. Nassikas tells us that space heating and cooling, which are responsible for perhaps a fifth of our primary energy consumption, must be focal points in our conservation efforts. Yet, he points out, most new buildings are still deficient in insulation and consideration of solar heat loads. Clearly, it doesn't make sense to indulge in different and expensive energy-increasing schemes and perhaps even rationing without taking corrective action in the area of building design. Mr. Nassikas suggests that tighter heating, cooling, and insulation standards are necessary. The Federal Housing Authority has upped its requirements for insulation in housing being built under its control. With full insulation of all buildings we might cut our heating fuel use by 20 percent!

Today, because home heaters are generally installed before homes are purchased, the builders make the decision on which type of heating to use. From the builder's point of view the logical choice is electric heat, since this is least expensive to install. But the difference shows up later on in considerably higher running costs—and greater over-all fuel use.

There is much that individuals can do also. Homeowners can lighten both heating and cooling loads by making sure that windows and doors are closed as airtight as possible. The judicious use of such devices as blinds and draperies can reduce heat load by as much as 50 percent; awnings, overhangs, and trees by 80 percent.

At times of high electricity use, changing the thermostat setting from 70° to 75° will reduce the air-conditioning load by 15 percent or more.* (People who have thermostatically controlled air conditioning tend to become careless in such matters.)

An important point to keep in mind is that, particularly in the case of electricity, a Btu saved at the point of use is equivalent to 3 or 4 saved at the point of generation.

* The U.S. Government Printing Office puts out two very useful booklets along this line: "11 ways to Reduce Energy Consumption and Increase Comfort in Household Cooling," and "7 Ways to Reduce Fuel Consumption in Household Heating."

Among changes we may well see in the future are greater use of solar heating, more in the way of centralized (hence more efficient) heating and cooling, less use of natural gas for electricity generation and more for home heating, and a shift from individual housing (with all those extra surfaces exposed to the elements) to apartments.

Consider too that the heat removed from refrigerators and freezers, and that given off by lights, adds to the heat load of air conditioners during the summer; and in the winter there's lots of free cold outside, yet we pour energy into our refrigerators and freezers which are battling to keep our food cold in our overheated homes. It should be possible, by placing these appliances near an outside wall and by use of some sort of duct arrangement, to dispose of the excess heat outdoors during the summer, or perhaps vent it into the water-heating system, while utilizing some of the excess cold outside during the winter.

Total Energy

In large installations, such as regional schools, hospitals, shopping malls, industrial plants, apartment complexes, and the like, it has been possible to design what have come to be called Total Energy plants. All of the various uses of energy are considered in the design of the plants that supply heating, cooling, electricity, and in some cases even lighting. By combining all of these a much greater efficiency of over-all use is possible. In general, Total Energy systems are fueled by oil or gas and some proportion of the heat that normally goes up the stacks is recovered and used for heating, cooling, and other special uses. Over-all thermal efficiencies can be as high as 65 percent.

The fuel most often used in Total Energy systems is gas,

and the high-temperature exhaust heat from the gas turbines is recovered in a boiler to generate steam, which is then used as the working fluid. Another line may carry chilled water, and return lines are required as well. When sections of a city are handled in this way, all the pipes can be laid in a single trench, though a thermal barrier may be necessary between the steam and cold water lines.

Total Energy plants saw rapid growth in the 1960's, and will probably see continued growth in the 1970's, at least partly because of improved, compact gas turbines being introduced. The concept may peak out in the mid- and late-1970's, however, because of natural gas supply problems. Whether synthetic gas can make up for this remains to be seen.

Total Utility?

Is Total Energy really total? There are already, as we have seen, refuse-disposal plants in which the heat produced is used for electrical generation; and there are sewage-disposal plants in which "digester gas" is used to fuel engines. The idea has thus been advanced of the Total Utility. This would encompass one huge energy utility which would take care of all the energy needs of a community—electric power, space and process heat, cooling, and waste disposal. By combining all of these into one efficient plant, energy needs can be provided for with the least amount of waste, and of course less air, water, and land pollution. Among the possible uses for the waste heat are the improvement of sewage processing (most chemical and biological processes go faster at higher temperatures) and production of pure water. Oyster "farmers" on Long Island are using waste heat from a nearby power plant to produce a large, succulent product in one-third the normal growing time. Other experiments are aimed at warming agricultural land to speed growing time, or to make it possible to grow warm-weather produce in cold areas.

Lighting

Of all our uses of energy, the production of light is surely the least efficient. Incandescent light, still used in most homes, gives off as light only about 5 to 10 percent of the energy fed into it. The rest goes off in invisible infrared or heat radiation. Fluorescent lighting does somewhat better, converting about 20 percent of the electricity it consumes to light. Perhaps 70 percent of the total illumination in the United States is now fluorescent, with incandescent and high-intensity lights (used on highways, for example, and more efficient than fluorescent lights) sharing the balance. The average efficiency of all our lighting is about 13 percent; when combined with the 33 percent efficiency of our electrical generation, the figure drops to 4 percent over-all—an appallingly low figure, especially since lighting uses almost a quarter of all the electricity generated, or 6 percent of the total energy budget.

We could right now cut down on use in various ways. Few benefit from office buildings that remain completely lit all or most of the night. Night workers and maintenance/cleaning personnel make some use of it, but mostly it's a kind of advertising. Advertising signs too use a lot of energy.

This brings us to the general question of night lighting. One of the biggest differences between our lives and those of our ancestors has been the fact that we have pushed away the night. Must we give up our "great white ways?" Our night-lighted bridges and monuments? Brightly lit streets? Much will depend on future supplies of energy. I for one feel it will be too bad if our bridges and monuments can no longer be lit up at night. There are also questions as to whether present standards for general illumination in commercial and industrial buildings are too high.

Another idea: There is usually plenty of light coming in through the enormous windows in modern office buildings, yet the lighting systems are designed to provide for peak requirements, meaning at night when there is no light coming in from outside. With the price of computers dropping drastically, we might consider variable perimeter lighting in which the amount of natural light coming in is sensed and the quantity of artificial light varied to keep total illumination constant.

Perhaps we might also find more efficient ways of producing artificial light. A fluorescent bulb might be designed that can be screwed into the many millions of incandescent light sockets. Or the even greater efficiencies of bioluminescence could be applied, if we could find out how biological creatures are able to produce light with practically no heat wastage. Once again the storage of energy comes to the fore, with the possibility of storing up light energy during day, to be emitted at night when needed.

Incandescent bulbs are lit by heat, and fluorescent bulbs by a combination of electricity and fluorescence (glowing of fluorescent powders). It has been suggested that "chemical pumping" (as is done in lasers) might offer a more efficient approach to production of light.

All in all it would seem that much could be done in the areas of heating, cooling, and lighting. Indeed, more can probably be done here than in conservation by means of less use of appliances. One calculation suggests that if *all* appliances, major and minor, were turned off, we would cut our use of electricity by only 11 percent, whereas buildings of all types account for more than half of all electricity used. (In New York City, reports Milton D. Rubin, a consulting scientist, 65 percent of the base electrical load is used for lighting!)

Something like 60 percent of the U.S. production of electrical energy is used by industry. Professor Berry estimates that with recycling, improvements in the basic methods of metal recovery and fabrication, building products to last 50 to 100 percent longer, and a general movement toward caring about energy expenditures, we could reduce our industrial needs for electrical energy by a factor of 10! And researchers at the University of California calculate that "it would be possible to reduce per capita energy consumption in the United States to 62 percent of current levels and maintain the same standard of living." *

* *Science News,* March 11, 1972, p. 171.

16

An Energy Policy for the Future

THE MAJOR PROBLEMS we face in the energy field can be summarized thus: availability, pollution, cost. The order of importance, particularly of the first two, depends upon who is making up the list.

We have looked into a number of possible solutions that may, or may not, give us answers to these problems. Mostly they have been technological solutions; and the ones that would seem, at the moment, best able to do the job—fusion and solar power—are just the ones that we are not sure can really be made to work (technically in the first case, and economically in the second).

In the meantime something has to be done. A myriad of decisions have to be made. Do we continue to increase our fuel consumption and do more damage to the environment? Do we go ahead with the Alaskan pipeline and continue to drill for offshore oil and gas? How about strip mining?

Can we find a way to produce more energy and do less damage to the environment? The only sure way to do this (if it can be done) is to increase our commitment to research.

Federal energy R & D funding proposed for fiscal year 1973, subject to approval by Congress

ITEM	Budget ($$10^6$)
Nuclear fission	356
Fossil fuels	136
Nuclear fusion	65
Solar energy	4
Geothermal energy	3
Related technologies	55
Total	622

(Source: *Science*, September 8, 1972, p. 876)

The $622 million shown in the table sounds like a lot of money, but amounts to only about .6 percent of total energy expenditures; this is far less than is spent in other industries. In the aerospace field some 16 percent of sales goes to research and development.

Do we import more fuel, thus exposing ourselves to the vagaries of international politics? Importing cars and sewing machines is one thing; importing energy is quite another. In the latter case the importer is truly at the mercy of the producer—which the producers have only recently begun to learn and take advantage of. Already some oil-rich countries have threatened to hold back on production until prices go up; have broken long-term agreements made only a few years or even months earlier, and have insisted on "participation" in running plants built by user countries on their soil. In some cases they have simply taken over ("nationalized") the industry.

There are other problems with importing, one of the most important of which is the cost in foreign exchange. Currently we import about one-quarter of our oil needs. At current rates of increase we may by 1980 be importing half our petroleum at a cost in foreign exchange of $15 to $20 billion a year; this would place a great strain on our already precarious balance of payments situation.

The Alaskan oil find will help, assuming we can get the oil down here. Similarly, increased offshore drilling would help. But as is well known, both approaches are under environmental fire.

American producers claim that if the government increases oil import quotas, this would hurt domestic production. (Domestic oil is more expensive than imports, even after transportation costs are included.) The projected availability of oil substitutes such as liquefied coal and oil shale sets a limit on the maximum cost of crude oil, domestic or imported. The producers' claim, which is accepted as valid, is that higher prices would bring forth more domestic oil, though clearly the more we use now the less we will have later.

On the other hand, prices on natural gas are limited by law, and are considered to be too low, which encourages greater use of this even more limited resource. (Congress is considering a change in this policy, however.)

Going Nuclear

The electric utility industry, in considering the present and future fossil fuels situation, and the cost of trying to meet increasingly strict air-quality standards, is, in general, "going nuclear." When a major Texas utility, right in the

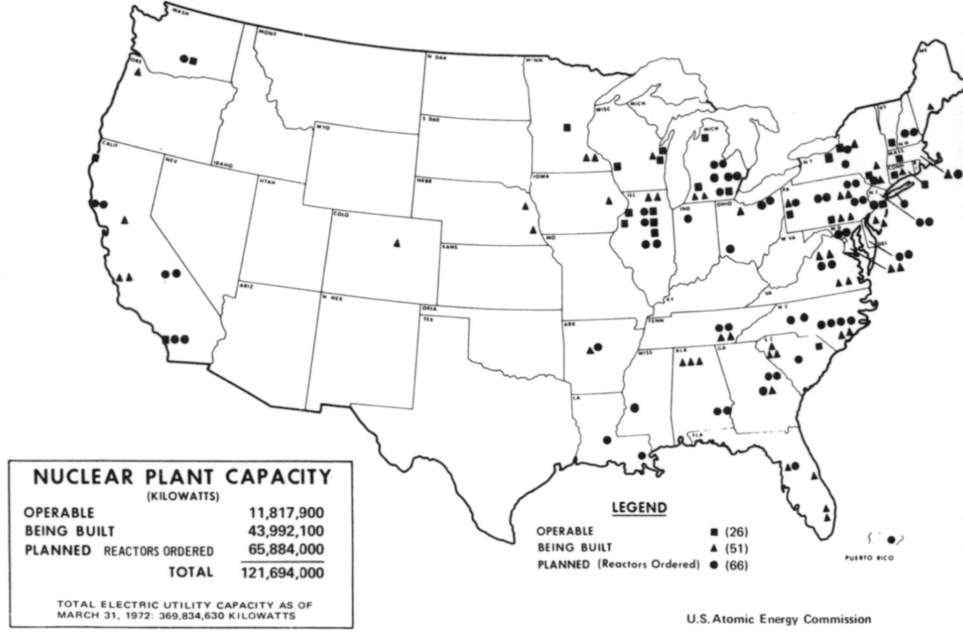

Nuclear power reactors in the U.S.

heart of gas country, orders a nuclear plant, we know a strong movement is in progress. Some 50 percent of power plant construction scheduled through 1980 is nuclear, and the figure rises to 75 percent for the 1980–90 period.

From the point of view of the consumer, how wise a decision is this? Again it depends on whom you listen to. The late Lev Artsimovitch, who was head of the Soviet fusion program, wrote, "Without the use of atomic energy, it is impossible to picture a continuously developing civilization on Earth." *

Clearly not everyone agrees, for nuclear plants have been

* *Bulletin of the Atomic Scientists,* October 1971, p. 50.

blocked right and left, for a wide variety of reasons. During all of 1971, not a single full-power operating license was granted by the Atomic Energy Commission. These delays are extremely costly, on the order of $2 to $4 million a month in carrying charges for each plant! Total cost for the industry is on the order of $5 to $6 billion for an average 12-month delay, which is not unusual. (By no means can all of this be put down to blockage by environmentalists; often poor planning, poor management, labor problems, and even technical problems are involved.)

Indeed, there are increasing feelings of unease regarding nuclear energy in general.* The sensible approach is to get all of the objections out in the open before the plant is built. (One of the problems, however, is the increasingly stringent laws that are being passed. The designers and builders are therefore often trying to "catch up with a moving target.")

In one happy case, Northern States Power (Minnesota) put together a task force of representatives from various interested parties regarding the siting of a planned power plant. The task force rejected the company's first three choices, but accepted the fourth. The power company decided to go along with the decision. Company management hopes in this way to prevent the kind of costly later delays that other companies have faced.

Perhaps this is the way to go, though a thorough shakeup in the licensing field would help. Utilities are now allowing 10 to 15 years from planning stage to completion. A single public hearing which used to take a day or two to complete may now run on for a year or more. "One-stop shopping" has been suggested as a licensing goal, but it may be harder to deal with a single powerful bureaucrat in a federal or regional office than with several local ones.

* Nevertheless, *The New York Times* (December 15, 1972) reported that 1972 was the most active ordering year in the history of the nuclear power industry.

In any case, something had better be done, or a prediction made by Glenn T. Seaborg, who until recently was chairman of the A.E.C., may come to pass. He said that if opponents to nuclear plants continue to prevail, "today's outcries about the environment will be nothing compared to the cries of angry citizens who find that power failures due to lack of sufficient generating capacity have plunged them into prolonged blackouts—not mere minutes, but hours, perhaps days—when their health and well-being and that of their families may be seriously endangered."

Endangered?

Is danger too strong a word? During a power outage in the summer of 1972, a seriously ill patient in a New York hospital, who was dependent upon the steady operation of a life-giving electrical machine, had to be maintained by the voluntary services of a group of firemen until power could be restored to the machine. Obviously there should have been backup sources of power, and in this case, ironically, workmen were in the process of installing some when the blackout occurred. But clearly there always will be situations in which it is not possible to provide backup power.

Let us say that nuclear energy is needed. Should we be making as strong a commitment as we are to the breeder? It has been suggested that this type of machine, with its large inventory of plutonium, may be even more dangerous than the conventional nuclear plant. Are we perhaps close enough to fusion that we should skip the breeder step altogether?

Perhaps. We have, nevertheless, made a strong commitment to the LMFBR. But so have Great Britain and the U.S.S.R. A number of workers in the field suggest that this is not necessarily because it is the best approach but simply because the technology in this area is most advanced. The gas-cooled and the molten-salt types, they say, offer even greater advantages, including a higher efficiency. Wouldn't it have made more sense for each of the various countries to have concentrated on a different type, instead of duplicating each other's research?

A final thought on the subject. In earlier chapters we considered the possibility of an all-electric economy. The question is: Even assuming that such an economy is based on admittedly clean and relatively safe fusion power, we must be careful not to make the same mistake we made when we moved toward the automobile as the basis for our transportation system. In other words, it seemed like a good idea at one time; today it is the cause of pollution, congestion, and a host of other major problems. Hopefully, we have learned something in the recent past and are better able to think ahead. We should do so.

As to the dangers inherent in nuclear energy, they do exist. As to how serious they are, more and better work needs to be done to find out, and now, before we get any deeper into it.

We should also keep the matter in some perspective; all of our technologies carry some danger with them. Bridge-, tunnel-, and skyscraper-building often take the lives of one or more workers; on May 13, 1972 three men were killed and three injured when a steam line burst in a Georgia power plant; carbon monoxide kills 1500 annually; 1000 lives are lost every year for the privilege and convenience of electric power; and that greatest killer of all, the automobile, takes 56,000 lives a year in the United States alone.

In less advanced societies, death from predatory animals, disease, and starvation are common. The idea of complete safety may simply be an impossible dream.

Energy Policy

To help us sort things out we need a more rational, equitable energy policy, instead of the hodgepodge of rules, laws, and regulations we have now. We have a Department of Transportation, of Housing and Urban Development. Why not a Department of Energy? Or, alternatively, why an *Atomic* Energy Commission? Perhaps it should be changed to just the Energy Commission. The closest we have come so far is the recent creation of an Energy Board under the Department of the Interior. In any case, we might then see the creation of a major national scientific laboratory devoted to energy research, something on the order of the National Accelerator Laboratory in Batavia, Illinois.

Most discussions of energy technology ask, somewhere along the line, "Is it economical?" or, "How does it compare in cost with present fossil or nuclear methods?" But this may not be the best, or even a sensible, approach. The development of any new technology is a long, complicated, expensive road. If the application of, say, solar energy is to become practicable in time for the period when it is needed, the time to begin a large-scale development effort is now. If we wait until the need becomes desperate, which is the usual pattern of large-scale development, the costs multiply by orders of magnitude.

Nor should we concentrate on any single type. It seems almost certain that we will need all types. Senator Mike McCormack foresees the following in energy development:

1980	Coal gasification
1985	Breeder
1985–90	Solar energy
2000	Satellite solar energy and fusion

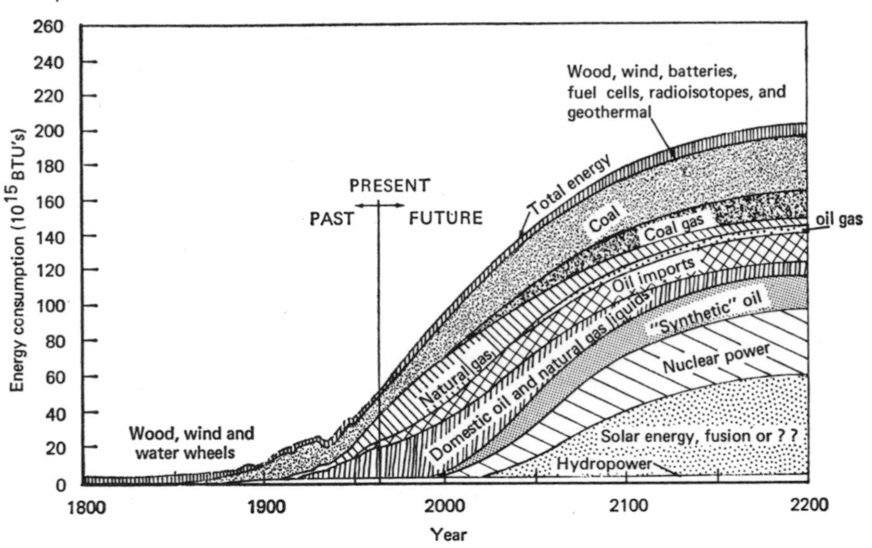

Energy consumption in the U.S., past, present, and projected.

Beyond the year 2100, Professor Earl Cook of Texas A & M University foresees United States and world needs being met by a combination of coal, nuclear power, and solar energy. The National Petroleum Council, on the other hand, feels that the most important new energy form in 1985 will be geothermal.

My own feeling is that all forms of energy production will find a place, though of course some, like tidal and wind energy, may be minimal.

Do We Need All That Energy?

It is important to keep in mind that we have come to depend on fuel and electric power for a large part of our modern existence. Whether this is right or wrong, good or bad, is at this point irrelevant. This is the way it is.

Even if the United States should attain (and maintain) what has come to be called Zero Population Growth, we would still end up with something like 75 million more Americans by the end of two generations, or roughly by the year 2030. (This is because of the large number of young people already born and moving into adulthood.)

Regarding per capita use of electricity and fuels, there are many in this richest of all countries who are still looking forward to a higher standard of living. B. I. Spinrad, senior physicist at Argonne National Laboratory, points out that "low-income America is still vastly underserved by electricity, with feebly illuminated dwellings and streets, and households with no major appliances except television being the rule." * Are those who have been working hard to be able to enjoy some of the luxuries offered by our industrialized society—air conditioning, motor boats, hi-fi sets, power mowers, a car!—are they to be told that there's an energy problem so we've had to limit production and use of items like these?

In searching out predictions of future use of energy in the United States, I have found the following. Out of the dozen or so I uncovered, not a single one sees total growth as being less than a steady 2 percent a year, and most range from 3 to 5 percent a year. (Present population growth is about 1 percent a year.) This means that future Americans are expected to be using 2 to 5 times as much energy per capita as they do now. Over-all, government forecasts see us consuming twice as much energy in 1985 as we do today.

This is the real world. We had better deal with it. For if we really do use up our fossil fuels, and if we do not find a way to produce clean energy, it could just happen that

* *Bulletin of the Atomic Scientists,* September 1971, p. 3.

the way of the East Indian peasant today will be the way of the American in the future.

On the other hand, world energy growth cannot continue to increase indefinitely. The only hope is some sort of steady-state system, meaning some maximum level of population and stabilization of rate of use. Clearly the only equitable way that this can come about is for all of mankind to attain some minimum standard of living. How, and when (or if) this will take place, and at what level it will be, no one knows now. Considering that 2 billion people in the world still do not have electricity, we see that we still have a long way to go.

Appendix

Conversion Factors

1 horsepower (hp)	= 745.7 watts
	= .7457 kilowatts (kw)
	= 42.4 British thermal units (BTUs) per minute
1 BTU	= 778.2 foot-pounds
	= 3.930×10^{-4} hp-hr
	= .2520 kilogram-calories
	= 2.930×10^{-4} kw-hr
1 kw-hr	= 3412 BTUs
	= 1.341 hp-hr
	= 3.6×10^{6} joules
	= 860 kg-calories
1 ev (electron-volt)	= 1.602×10^{-12} erg
	= 15.187×10^{-23} BTUs

Bibliography

BOOKLETS, REPORTS, AND PAPERS

Anvanced Nonthermally Polluting Gas Turbines in Utility Applications, 1971 (available from U.S. Government Printing Office).

American Oil Shale Corporation and the Atomic Energy Commission (Chapin, C. E., et al.), "A Feasibility Study of Plowshare Geothermal Power Plant," April 1971.

Anderson, L. L., "Energy Potential from Organic Wastes: A Review of the Quantities and Sources," Bureau of Mines, 1972.

Atomic Energy Commission, "Forecast of Growth of Nuclear Power, January 1971 (available from U.S. Government Printing Office).

Battelle Memorial Institute, "A Review and Comparison of Selected United States Energy Forecasts," prepared for the Executive Office of the President, December 1969 (available from the U.S. Government Printing Office).

Corliss, W. R., "Direct Conversion of Energy," U.S. Atomic Energy Commission, 1964.

———, "Nuclear Reactors for Space Power," U.S. Atomic Energy Commission, 1971.

———, and R. L. Mead, "Power from Radioisotopes," U.S. Atomic Energy Commission, 1971.

Council on Environmental Quality, "Environmental Quality, The First Annual Report," August 1970 (available from the U.S. Government Printing Office).

Departments of the Interior and Agriculture, "Environmental Criteria for Electric Transmission Systems," October 1970.

Division of Operations Analysis and Forecasting, "Forecast of Growth of Nuclear Power," U.S. Atomic Energy Commission, January 1971.

Division of Technical Information, "Nuclear Power and the Environment," U.S. Atomic Energy Commission, 1969.

Dukert, J. M., "Thorium and the Third Fuel," U.S. Atomic Energy Commission, 1970.

Eastlund, B. J., and W. C. Gough, "The Fusion Torch, Closing the Cycle from Use to Reuse," U.S. Atomic Energy Commission, May 15, 1969 (available from the U.S. Government Printing Office).

Electric Power Survey Committee, "50th Semi-Annual Electric Power Survey," Edison Electric Institute, October 1971.

"Energy: Transactions in Time," Special Issue of *Kaiser News*, Kaiser Aluminum and Chemical Co., 1970.

Environmental Policy Division, "The Economy, Energy, and the Environment," Joint Economic Committee, Congress of the United States, September 1, 1970.

Federal Power Commission, "Electric Power Transmission and the Environment," undated.

"Hydroelectric Power Resources of the United States, Developed and Undeveloped," U.S. Government Printing Office, 1968.

Federal Power Commission, "World Power Data," U.S. Government Printing Office, 1968.

Fu, Y. C., et al., "Conversion of Bovine Manure to Oil," paper presented at 164th National Meeting, American Chemical Society, Division of Fuel Chemistry, August 28, 1972 (contained in Vol. 17, No. 1, *General Papers . . .*).

Glasstone, Samuel, "Controlled Nuclear Fusion," U.S. Atomic Energy Commission, 1965.

Gough, W. C., "Why Fusion?" U.S. Atomic Energy Commission, June 1970 (available from U.S. Government Printing Office).

———, and B. J. Eastlund, "Energy, Wastes and the Fusion Torch," U.S. Atomic Energy Commission, April 27, 1971.

Gould, R. W., et al., "Progress in Controlled Thermonuclear Research," U.S. Atomic Energy Commission, December 1970 (available from the U.S. Government Printing Office).

Hogerton, J. F., "Atomic Fuel," U.S. Atomic Energy Commission, November 1963.

Hogerton, J. F., "Nuclear Reactors," U.S. Atomic Energy Commission, 1970.

House Committee on Interior and Insular Affairs, Fuel and Energy Resources, 1972, Part 1 (hearings held on April 10-13, 1973; available from the U.S. Government Printing Office).

IEEE Power Engineering Society, "The Great Environmental Debate and the Power Industry," 1971 (available from the Institue of Electrical and Electronic Engineers).

Joint Committee on Atomic Energy, "Controlled Thermonuclear Research," Part 1: Hearings Before the Sub-Committee on Research, Development, and Radiation; Part 2: Appendixes 1972 (available from the U.S. Government Printing Office).

Joint Committee on Atomic Energy (Hearings), "Liquid Metal Fast Breeder Reactor (LMFBR) Demonstration Plant," Congress of the United States, 1972 (available from the U.S. Government Printing Office).

Joint Committee on Atomic Energy (Hearings), "Nuclear Power and Related Energy Problems –1968–1970," Congress of the United States, December 1971 (available from the U.S. Government Printing Office).

Lyerly, R. L., and W. Mitchell, "Nuclear Power Plants," U.S. Atomic Energy Commission, 1969.

Mitchell, W., III, and S. E. Turner, "Breeder Reactors," U.S. Atomic Energy Commission, 1971.

Meredith, D. L., "Nuclear Power Plant Siting: A Handbook for the Layman," Marine Advisory Service, University of Rhode Island, June 1972.

Morrison, W. E., and C. L. Reading, *An Energy Model for the United States, Featuring Energy Balances for the Years 1947 to 1965 and Projections and Forecasts to the Years 1980 and 2000.* U.S. Department of Interior, Bureau of Mines, No. 8384, 1968.

National Electric Reliability Council, Impact of a 12-Month Delay of New Nuclear and Fossil-Fired Steam Generating Plants on the Adequacy of Electric Power Supply in the United States, February 1972.

"National Gas Supply and Demand, 1971–1990," U.S. Government Printing Office, 1972.

National Petroleum Council, "U.S. Energy Outlook, An Initial Appraisal 1971–1985," Vol. 1, July 1971; Vol. 2, Summaries of Task Group Reports, November 1971.

National Power Survey Task Force on Environment, "Managing the Power Supply and the Environment," Federal Power Commission, July 1, 1971.

Office of Information Services, "Nuclear Terms, A Glossary," 2nd ed., U.S. Atomic Energy Commission, undated.

Pittman, F. K., "Plan for the Management of AEC-Generated Radioactive Wastes," U.S. Government Printing Office, January 1972.

R & D Goals Task Force, "Electric Utilities Industry Research and Development Goals Through the Year 2000," Electric Research Council, June 1971.

Rosa, R. J., "MHD Power Generation: Basic Principles and State of the Art," Avco Everett Research Laboratory, March 1972.

Singleton, A. L., "Sources of Nuclear Fuel," U.S. Atomic Energy Commission.

Summers, C. M., "Electrical Energy by Direct Conversion," No. 147, The Office of Engineering

Research, Oklahoma State University, March 1966.

Thompson, H. E., *"Solar Houses and Solar House Models,"* Edmund Scientific Co., 1972.

U.S. Department of the Interior, "United States Energy, A Summary Review," January 1972.

U.S. Department of Interior, "United States Petroleum Through 1980."

Zareski, G. K., "The Gas Supplies of the United States—Present and Future," paper presented at 164th National Meeting American Chemical Society, Division of Fuel Chemistry, August 28, 1972 (contained in Vol. 17, No. 1 *General Papers . . .*).

ARTICLES

Aaronson, T., "The Black Box" (fuel cell), *Environment,* December 1971.

Abelson, P. H., "Scarcity of Energy," *Science,* September 25, 1970.

"AEC Causes Anger and Delight," *Nature,* September 1971.

Agarwal, P. D., "Electricity Not Such a Clean Fuel," *Automotive Engineering,* February 1971.

Alfven, Hannes, "Energy and Environment," *Bulletin of the Atomic Scientists,* May 1972.

Alfven, Hannes, "Spacecraft Propulsion: New Methods," *Science,* April 14, 1972.

"Another SST?" (re dangers of the liquid-metal-cooled fast breeder reactor) *Environment,* July–August 1971

Armagnac, A. P., "Hot and Cold Running Air," *Popular Science,* July 1969.

———, "Power from Crystals to Drive Strange New Tools," *Popular Science,* August 1969.

Asimov, Isaac, "How Breeder Reactors Work," *Science Digest,* January 1966.

Bagge, C. E., "Coal, An Overlooked Energy Source," *Vital Speeches of the Day,* April 1, 1972.

Barnea, J., "Geothermal Power," *Scientific American,* January 1972.

Benedict, M., "Electric Power From Nuclear Fission," *Bulletin of the Atomic Scientists,* September, 1971. (Also in *Technology Review,* October–November 1971.)

Bengelsdorf, I., "Are We Running Out of Fuel?" *National Wildlife,* February 1971.

Bennett, R. R., "Planning for Power—a Look at Tomorrow's Power Stations," *IEEE Spectrum,* September 1968.

Benford, J., and G. Benford, "Intense Electron Beams—A Fusion Match?" *New Scientist,* November 30, 1972.

Berry, R. S., "Recycling, Thermodynamics and Environmental Thrift," *Bulletin of the Atomic Scientists,* May 1972 (technical, but interesting).

Blumer, M., et al., "A Small Oil Spill," *Environment,* March 1971.

Bibliography

Boersma, L. L., "Nuclear Waste Heat Could Turn Fields into Hotbeds," *Crops and Soils,* April–May 1970.

Bohn, H. L., "A Clean New Gas," *Environment,* December 1971.

———, "Methane From Waste," *Environmental Science and Technology,* July 1971.

Bowen, R. G., and E. A. Groh, "Geothermal–Earth's Primordial Energy," *Technology Review,* October–November 1971.

Breckenfeld, G., "How the Arabs Changed the Oil Business," *Fortune,* August 1971.

Bruckner, A. II, W. J. Fabrycky, and J. E. Shamblin, "Economic Optimization of Energy Conversion with Storage," *IEEE Spectrum,* April 1968.

Bump, T. R., "Third Generation Breeder Reactors," *Scientific American,* May 1967.

Burck, Gilbert, "The FPC is Backing Away From the Wellhead," *Fortune,* November 1972 (re deregulation of wellhead prices of gas).

"Carbon Monoxide Wrings Oil From Lignite on Garbage," *Chemical Engineering,* October 1971.

Chapman, D., et al., "Electricity Demand Growth and the Energy Crisis," *Science,* 17 November 1972.

Charlier, R. H., "Tidal Energy" *Sea Frontiers,* November–December 1969.

Clark, W., "How to Harness Sunpower and Avoid Pollution," *Smithsonian,* November 1971.

Cole, D. E., "The Wankel Engine," *Scientific American,* August 1972.

Cook, Earl, "Energy for Millenium Three," *Technology Review,* December 1972.

———, "Energy Sources for the Future," *The Futurist,* August 1972.

———, "The Flow of Energy in an Industial Society," *Scientific American,* September 1971.

Coppi, B., and J. Rem, "The Tokamak Approach in Fusion Research," *Scientific American,* July 1972.

Culler, F. L., Jr., and W. O. Harms, "Energy from Breeder Reactors," *Physics Today,* May 1972.

Dadisman, Quincy, "Arctic Pipeline: Known Problems and Unknown Effects," *The Nation,* October 2, 1972.

Driscoll, M. J., "The Role of Nuclear Power in Achieving the World We Want," *The Science Teacher,* November 1970.

Dyson, F. J., "Energy in the Universe," *Scientific American,* September 1971.

Ehricke, K. A., "Extraterrestrial Imperative," *Bulletin of the Atomic Scientists,* November 1971.

Ehrlich, P. R., and J. P. Holdren, "The Energy Crisis," *Saturday Review,* August 7, 1971.

———, "Impact of Population Growth," *Science,* March 26, 1971.

Eliassen, R., "Power Generation and the Environment," *Bulletin of the Atomic Scientists,* September 1971.

Emelyanov, V. S., "Nuclear Energy in the Soviet Union," *Bulletin of the Atomic Scientists,* November 1971.

"Energy for the World's Technology *Review,* October–November 13, 1969.

Faltermayer, Edmund, "The Energy 'Joyride' Is Over," *Fortune,* September 1972.

Farmer, F. R., "Safety and Nuclear Power Plants: A British View," *Bulletin of the Atomic Scientists,* November 1971.

Feirtag, Michael, "65 Cars in Search of the Future," *Technology Review,* October–November 1972.

Fenner, D., and J. Klarmann, "Power From the Earth," *Environment,* December 1971.

Finneran, J. A., "SNG—Where Will It Come From, and How Much Will It Cost?" *The Oil and Gas Journal,* July 17, 1972.

Flood, H. S., and B. S. Lee, "Fluidization," *Scientific American,* July 1968 (re coal combustion).

Free, J. R., "Cryogenic Power Lines: Cool Aid for Our Energy Crisis," *Popular Science,* October 1972.

Friedlander, G. D., "The Story of Bonneville Power: 1937–1968–1987. Dams of the Columbia River Basin," *IEEE Spectrum,* November 1968 and December 1968.

Frysinger, G. R., "Energy Conversion Devices Today," *Research/Development,* March 1969.

Gannon, Robert, "What Are Our Alternatives?" *Science Digest,* December 1971.

Gates, D. M., "The Flow of Energy in the Biosphere," *Scientific American,* September 1971.

Geesaman, D. P., "Plutonium and the Energy Decision," *Bulletin of the Atomic Scientists,* September 1971.

Gendlin, F., "The Palisades Protest: A Pattern of Citizen Intervention," *Bulletin of the Atomic Scientists,* November 1971.

Giocoletto, L. J., "Energy Storage and Conversion," *IEEE Spectrum,* February 1965.

Gillette, R., "Nuclear Reactor Safety: A New Dilemma for the AEC," *Science,* July 9, 1971.

———, "Schlesinger and the AEC: New Sources of Energy," *Science,* January 14, 1972.

Gilluly, R. H., "A Home on the Range for a Vast Industry," *Science News,* March 4, 1972 (re power generation complex).

———, "Nitrogen Oxides, Autos and Power Plants," *Science News,* April 15, 1972.

Glaser, P. E., "Power from the Sun: Its Future," *Science,* November 22, 1968.

"Gobar Gas: Methane Experiments in India," *The Mother Earth News,* November 1971.

Gofman, J. W., "Nuclear Power and Ecocide: An Adversary View of New Technology," *Bulletin of the Atomic Scientists,* September 1971.

Gough, W. C., and E. J. Eastlund, "The Prospects of Fusion Power," *Scientific American,* February 1971.

Goldman, M. I., "The Convergence of Environmental Disruption," *Science,* October 2, 1970. (The U.S.S.R. has the same problems we have.)

Graham, F., Jr., "Tempest in a Nuclear Teapot," *Audubon,* March 1970.

Gravel, Mike, "Gentle Solutions for Our Energy Needs," *Congressional Record,* February 9, 1972, Part II.

Gregory, D. P., "The Hydrogen Economy," *Scientific American,* January 1972.

Grimmer, D. P. and K. Luszczynski, "Lost Power," *Environment,* April 1972 (re transportation).

Gross, E. T. B., "Efficiency of Thermoelectric Devices," *American Journal of Physics,* November 1961.

Halacy, D. S., Jr., "The Solar Alternative to Atomic Power," *Science Digest,* March 1972.

Hamilton, L. D., "On Radiation Standards," *Bulletin of the Atomic Scientists,* March 1972.

Hammond, A. L., "The Fast Breeder Reactor: Signs of a Critical Reaction," *Science,* April 28, 1972.

———, "Fission: The Pro's and Con's of Nuclear Power," *Science,* October 13, 1972.

———, "Solar Energy: A Feasible Source of Power?" *Science,* May 14, 1971.

Hannon, B. M., "Bottles, Cans, Energy," *Environment,* March 1972.

Hardesty, C. H., Jr., "The Critical Path to Adequate Supplies of Energy," *Vital Speeches of the Day,* July 1, 1972.

Hawkins, W. J., "Pot Gets Hot While Stove Keeps Cool," *Popular Science,* April 1972.

Head, J. W., "Nature's Hot Water Heats Greenhouses," *Farm Electrification,* September–October, 1970.

Hirst, E., and J. C. Moyers, "Efficiency of Energy Use in the United States," *Science,* March 30, 1973.

Hogerton, J. F., "The Arrival of Nuclear Power," *Scientific Amercan,* February 1968.

Hohenemser, K., and J. McCaull, "The Windup Car," *Environment,* June 1970.

Holcomb, R. W., "Power Generation: The Next 30 Years," *Science,* January 9, 1970.

"Horizon to Horizon," *Environment,* March 1971 (review of major oil spills).

Hubbert, M. K., "Energy Resources," Chapter 8 in *Resources and Man,* National Academy of Sciences–National Research Council, W. H. Freeman and Co., 1969.

———, "The Energy Resources of the Earth," *Scientific American,* September 1971.

Ikard, F. N., "Energy and Economics," *Vital Speeches of the Day,* September 15, 1972.

———, "The Energy Gap: A Search for Solutions," *Petroleum Today,* Winter 1971 (American Petroleum Institute).

Inglis, D., "Nuclear Energy and the Malthusian Dilemma," *Bulletin of the Atomic Scientists,* February 1971.

Jacobsen, Sally, "Turning Up the Gas: AEC Prepares Another Nuclear Gas Stimulation Shot," *Bulletin of the Atomic Scientists,* May 1972.

Jensen, Albert C., "Fish and Power Plants," *Conservationist,* December 1969–January 1970.

Jones, L. W., "Hydrogen: A Fuel to Run Our Engines in Clean Air," *Saturday Evening Post,* Spring 1972.

———, "Liquid Hydrogen as a Fuel for the Future," *Science,* October 22, 1971.

Kentfield, Calvin, "New Showdown in the West," *New York Times Magazine,* January 28, 1973 (re strip mining).

Kovarik, T. J., "Raped Earth of Strip-Miners—Can It Be Healed?" *Science Digest,* April 1972.

Kusko, Alexander, "A Prediction of Power System Development," *IEEE Spectrum,* April 1968.

Lapp, R. E., "The Four Big Fears About Nuclear Power," *New York Times Magazine,* February 7, 1971.

Lebedev, B. P., and S. S. Rokotian, "E H V Transmission in the U.S.S.R. Power Grid," *IEEE Spectrum,* February 1967.

Lederberg, J., "Squaring an Infinite Circle: Radiobiology and the Value of Life," *Bulletin of the Atomic Scientists,* September 1971.

Lessing, Lawrence, "The Coming Hydrogen Economy," *Fortune,* November 1972.

———, "New Ways to More Power with Less Pollution," *Fortune,* November 1970.

———, "Power from the Earth's Own Heat," *Fortune,* June 1969.

Levin, Franklyn, "Oil Exploration Technology," *Science and Technology,* February 1969.

Lewis, W. B., and A. M. Marko, "Seeking a Convergent View: The Unfinished Detective Story," *Bulletin of the Atomic Scientists,* November 1971 (re nuclear power).

Lidsky, L. M., "The Quest for Fusion Power," *Technology Review,* January 1972.

Likens, G. E., "Acid Rain," *Environment,* March 1972.

Lindop, P. J., and J. Rotblat, "Radiation Pollution of the Environment," *Bulletin of the Atomic Scientists,* September 1971.

Lindsley, E. F., "Battery-Powered Yard Tracs Are on the Way," *Popular Science,* August 1969.

Loving, Rush, Jr., "A Vast New Storehouse for Electricity," *Fortune,* December 1971.

Lubin, M. J., and A. P. Fraas, "Fusion by Laser," *Scientific American,* June 1971.

Lubkin, G. B., "AEC Opens Up on Laser-Fusion Implosion Concept," *Physics Today,* August 1972.

Katz, Milton, "Decision-Making in the Production of Power," *Scientific American,* September 1971.

Luten, D. B., "The Economic Geography of Energy," *Scientific American,* September 1971.

Lynch, C. J., "Emerging Power Sources," *Science and Technology,* October 1967.

Makhijani, A. B., and A. J. Lichtenberg, "Energy and Well Being," *Environment,* June 1972.

Maugh, T. H., II, "Fuel Cells: Dispersed Generation of Electricity," *Science,* December 22, 1972.

———, "Fuel From Wastes: A Minor Energy Source," *Science,* November 10, 1972.

———, "Gasification: A Rediscovered Source of Clean Fuel," *Science,* October 6, 1972.

Mayer, L. A., "Why the U.S. Is in an Energy Crisis," *Fortune,* November 1970.

McCaull, J., "A Lift for the Auto," *Environment,* December 1971 (re flywheel).

———, "Windmills," *Environment,* February 1973.

McDonald, J., "Oil and the Environment: The View from Maine," *Fortune,* April 1971.

Meinel, A. B. and M. P. Meinel, "Physics Looks at Solar Energy," *Physics Today,* February 1972.

————, "Is It Time for a New Look at Solar Energy?" *Bulletin of the Atomic Scientists,* October 1971.

Metz, W. D., "New Means of Transmitting Electricity: A Three-Way Race," *Science,* December 1, 1972.

———, "Power Gas and Combined Cycles: Clean Power from Fossil Fuels," *Science,* January 5, 1973.

Mills, R. G., "The Promise of Controlled Fusion," *IEEE Spectrum,* November 1971.

Moss, L. I., "Taxing U.S. Polluters," *Saturday Review,* August 7, 1971.

"The Next Step Is the Breeder Reactor," *Fortune,* March 1967.

"Nuclear Power Goes Critical," *Fortune,* March 1967.

Perelman, M. J., "Farming with Petroleum," *Environment,* October 1972.

Perry, Harry, "The Energy Crisis," Chapter in the *1972 Britannica Yearbook of Science and the Future,* 1972.

Rocks, L., and R. P. Renyon, *The Energy Crisis,* Crown Publishers, 1973.

Rose, D. J., "Controlled Nuclear Fusion: Status and Outlook," *Science,* May 21, 1971.

Sagan, L. A., "Human Costs of Nuclear Power," *Science,* August 11, 1972.

Satterfield, C. N., "Nitrogen Oxides: A Subtle Control Task," *Technology Review,* October–November 1972.

Schurr, S. H., "Energy," *Scientific American,* September 1963.

Seaborg, G. T., "On Misunderstanding the Atom," *Bulletin of the Atomic Scientists,* September 1971.

———, "For a U.S. Energy Agency," *Science,* June 16, 1972.

Seaborg, G. T., and J. L. Bloom, "Fast Breeder Reactors," *Scientific American,* November 1970.

Shachtman, T., "Getting More Power to the People," *Ecology Today,* September 1971.

Sherman, Don, "Gosh Mr. Wizard, Is *That* How It Works?" *Car and Driver,* December 1972 (alternatives to the internal combustion engine).

Slappey, S. G., "Heading Off an Energy Crisis," *Nation's Business,* July 1971.

Smock, R. W., "Gas from Coal Is on Utility Back Burner," *Electric Light and Power,* February 1972.

Snowden, Donald P., "Superconductors for Power Transmission," *Scientific American,* April 1972.

Spinrad, B. I., "America's Energy Crisis: Reality or Hysteria?" *Bulletin of the Atomic Scientists,* September 1971.

Squires, A. M., "Clean Power from Dirty Fuels," *Scientific American,* October 1972.

———, "Clean Power from Coal," *Science,* August 28, 1970.

Starr, Chauncey, "Energy and Power," *Scientific American,* September 1971.

Sternglass, E. J., "The Health Effects from Radiation," *Vital Speeches of the Day,* September 1, 1971.

Stong, C. L., "How to Make an Electrochemical Cell," *Scientific American,* November 1967.

Stonier, Tom, "An International Solar Energy Development Decade," *Bulletin of the Atomic Scientists,* May 1972.

Suica, J. S., "Power Development in Yugoslavia," *Bulletin of the Atomic Scientists,* November 1971.

Summers, C. M., "The Conversion of Energy," *Scientific American,* September 1971.

Swinkels, D. A. J., "Electrochemical Vehicle Power Plants," *IEEE Spectrum,* May 1968.

Tamplin, A. R., "Issues in the Radiation Controversy," *Bulletin of the Atomic Scientists,* September 1971.

Teller, Edward, "Can We Harness Nuclear Fusion?" *Popular Science,* May 1972.

Thomsen, D. E., "Farming the Sun's Energy," *Science News,* April 8, 1972.

———, "MHD: High Promise, Unsolved Problems," *Science News,* August 26, 1972.

———, "Mirror, Mirror on the Wall, Which Is the Fairest Plasma of Them All?" *Science News,* January 8, 1972.

Till, J. E., "Science and Politics in the Controversy Over Nuclear Power Hazards," *Science Forum,* August 1971.

Trotter, R. J., "Is Hydrogen the Fuel of the Future?" *Science News,* July 15, 1972.

Turner, Paul, "The Radiation Controversy," *Vital Speeches of the Day,* September 1, 1971.

Vollmer, J., and D. Gandolfo, "Microsonics," *Science,* January 14, 1972.

Walsh, E. M., "Electrogasdynamic Energy Conversion" *IEEE Spectrum,* December 1967.

Weaver, K. F., "Search for Tomorrow's Power," *National Geographic,* November 1972.

Weinberg, A. M., "Social Institutions and Nuclear Energy," *Science,* July 7, 1972.

Weinberg, A. M., and R. P. Hammond, "Global Effects of Increased Use of Energy," *Bulletin of the Atomic Scientists,* March 1972.

Wilford, John Noble, "Nation's Energy Crisis," *The New York Times,* July 6–8, 1971.

Williams, R. H., "When the [Gas] Well Runs Dry," *Environment,* June 1972.

Wilson, R., "Power Policy—Plan or Panic?" *Bulletin of the Atomic Scientists,* May 1972 (a prescription for action).

Wolff, A., "The Price of Power," *Harper's,* May 1972.

Wood, L., and J. Nuckolls, "Fusion Power," *Environment,* May 1972.

Zener, Clarence, "Solar Sea Power," *Physics Today,* January 1973.

BOOKS

Angrist, S. W., *Direct Energy Conversion,* 2nd ed., Allyn and Bacon, 1971.

Barnaby, Frank, *Man and the Atom: The Uses of Nuclear Energy,* Funk and Wagnalls, 1972.

de Bell, Garrett, ed., *The Environmental Handbook,* Ballantine, 1970.

Chalmers, B., *Energy,* Academic Press, 1963.

Curtis, R. and E. Hogan, *Perils of the Peaceful Atom,* Ballantine, 1969.

Daniels, F., *Direct Use of the Sun's Energy,* Yale University Press, 1964.

Dicke, R. H., *Gravitation and the Universe: Jayne Memorial Lecture for 1969,* American Philosophical Society, 1970.

Efford, I. E., and B. M. Smith, eds., *Energy and the Environment,* University of British Columbia Institute of Resource Ecology, 1972.

Emmerich, W., et al., *Energy Does Matter,* Walker, 1964.

The Energy Policy Staff, Office of Science and Technology, *Electric Power and the Environment,* U.S. Government Printing Office, 1970.

Environmental Policy Division, Legislative Reference Service, Library of Congress, *The Economy, Energy, and the Environment: A Background Study Prepared for the Use of the Joint Economic Committee, Congress of the United States,* U.S. Government Printing Office, 1970.

Esposito, J. C., *Vanishing Air,* Grossman, 1970.

Fabricant, Neil, and R. M. Hallman, *Toward a Rational Power Policy, Energy, Politics, and Pollution,* Braziller, 1971.

Federal Power Commission, *Statistics of Privately Owned Electric Utilities in the United States, 1969,* U.S. Government Printing Office, 1971.

Glasstone, Samuel, *Sourcebook on*

Atomic Energy, Van Nostrand, 1967.

Gofman, J. W., and A. R. Tamplin, *Poisoned Power: The Case Against Nuclear Power Plants,* Rodale Press, 1971.

Grey, Jerry, *The Race for Electric Power,* Westminster Press, 1972.

Guyol, N. B., *The World Electric Power Industry,* University of California Press, 1969.

Halacy, D. S., Jr., *The Coming Age of Solar Energy,* Harper & Row, revised edition, 1973.

Hammond, A., et al., *Energy and the Future,* American Association for the Advancement of Science, 1973 (proceedings of a conference).

Harrison, G. R., *The Conquest of Energy,* Morrow, 1968.

Holdren, John, and P. Herrera, *Energy: A Crisis in Power,* Sierra Club (distr. Scribners), 1971.

Hottel, H. C. and J. B. Howard, *New Energy Technology: Some Facts and Assessments,* MIT Press, 1971.

Jensen, W. G., *Energy and the Economy of Nations,* Transatlantic, 1970.

Kelly, Elizabeth, ed., *Ultrasonic Energy: Biological Investigations and Medical Applications,* University of Illinois Press, 1965.

Landsberg, H. H., and S. H. Schorr, *Energy in the United States: Sources, Uses, and Policy Issues,* Random House, 1968.

Landsberg, H. S., *Natural Resources for U.S. Growth; A Look Ahead to the Year 2000,* Johns Hopkins, 1964, 1967.

Levine, S. N., ed., *Selected Papers on New Techniques for Energy Conversion,* Dover, 1961 (for the scientist or science historian only).

Manners, Gerald, *The Geography of Energy,* Hillary House, 1964.

Marx, W., *Oilspill,* Sierra Club (distr. Scribners), 1971.

Morowitz, H. J., *Energy Flow in Biology,* Academic Press, 1968.

Mott-Smith, Morton, *The Concept of Energy Simply Explained,* Dover, 1964.

Murdock, W. W., ed., *Environment: Resources, Pollution and Society,* Sinauer Associates, 1971.

National Academy of Sciences–National Research Council Committee on Resources and Man, *Resources and Man: A Study and Recommendations,* 1969.

National Economic Research Associates, Inc., *Energy Consumption and Gross National Product in the United States: An Examination of a Recent Change in the Relationship,* 1971.

Odom, H. T., *Environment, Power and Society,* Wiley-Interscience, 1971.

Phillipson, J., *Ecological Energetics,* Edward Arnold, 1966.

Reynolds, John, *Windmills and Watermills,* Praeger, 1970.

Ridgway, James, *The Last Play: The Struggle to Monopolize the World's Energy Resources,* Dutton, 1973.

Rocks, L., and R. P. Runyon, *The Energy Crisis,* Crown, 1972.

Russel, C. R., *Elements of Energy Conversion,* Pergamon, 1967.

Saltonstall, D., Jr., and J. K. Page, Jr., *Brown-Out and Slow-Down,* Walker, 1972.

Seaborg, G. T., (speeches) *"Peaceful Uses of Nuclear Energy,"* U.S. Atomic Energy Commission, July 1970 (free).

———, and W. Corliss, *Man and Atom,* Dutton, 1971.

Sherfield, Lord, ed., *Economic and Social Consequences of Nuclear Energy,* Oxford University Press, 1972.

Sparrow, W. J., *Count Rumford of Woburn, Mass.,* Crowell, 1964.

Stokley, James, *The New World of the Atom,* Ives Washburn, 1970 (rev. ed.).

Talbot, A. R., *Power Along the Hudson: The Storm King Case and the Birth of Environmentalism,* Dutton, 1972.

Williamson, H. F., et al., The American Petroleum Industry, Vol. 1, *The Age of Illumination,* 1859–1899 (pub. 1959); Vol. 2, *The Age of Energy,* 1899–1959 (pub. 1963), Northwestern University Press.

Technology Review, Energy to the Year 2000, MIT, 1972 (Articles from *Technology Review,* October–November 1971, December 1971, and January 1972.)

Wilson, Mitchell, and the Editors of *Life, Energy,* 1963.

Work Group on Energy Products, *Man's Impact on the Global Environment: Assessment and Recommendations for Action,* MIT Press, 1970.

Young, L. B., *Power Over People,* Oxford University Press, 1973.

Zarem, A. M., and D. D. Erway, eds., *Introduction to the Utilization of Solar Energy,* McGraw-Hill, 1963.

Index

PICTURE CREDITS

Page	
25	"Why Fusion?" by W. C. Gough
27	National Petroleum Council, *U.S. Energy Outlook: An Initial Appraisal 1971–1985*, November 1971
36	"Nuclear Reactors for Space Power," Atomic Energy Commission
39	Avco Everett Research Laboratory
41	Avco Everett Research Labortory
44	British Petroleum
48	M. King Hubbert/*Resources and Man,* publication 1703, Committee on Resources and Man, National Academy of Sciences-National Research Council, W. H. Freeman and Co., San Francisco, 1969
51	M. King Hubbert/*Resources and Man,* publication 1703, Committee on Resources and Man, National Academy of Sciences-National Research Council, W. H. Freeman and Co., San Francisco, 1969
61	Gulf General Atomic
63	U.S. Atomic Energy Commission
70	Curtiss-Wright
77	Institute of Gas Technology
89	N. J. Public Service Electric and Gas Company

Page	
96	Pacific Gas and Electric Company
100	United Nations
104	*Physics Today*
110	California Institute of Technology
112	*Physics Today*
113	*Physics Today*
123	Princeton University
135	Princeton University Plasma Physics Laboratory
137	Princeton University Plasma Physics Laboratory
138	Lawrence Radiation Laboratory
139	Lawrence Radiation Laboratory
141	U.S. Atomic Energy Commission
155	Oklahoma State University/ R. J. Schoeppel
164	The Garrett Corporation
169	Su Kemper, McKee Engineering Corporation
173	Pratt and Whitney Aircraft
184	Brooklyn Union Gas Company
188	General Electric Company
193	U.S. Atomic Energy Commission
196	"Why Fusion?" by W. C. Gough
210	U.S. Atomic Energy Commission
215	L. P. Gaucher, *Journal of the Solar Energy Society,* 1965

BELMONT COLLEGE LIBRARY